THERMISTORS

Thermistors

F. J. HYDE, D.SC., M.SC., B.SC.
Late Professor and Head of the Department of
Electrical and Electronic Engineering
Royal Military College of Science
Shrivenham. Swindon, Wiltshire.

LONDON
ILIFFE BOOKS

THE BUTTERWORTH GROUP

ENGLAND

Butterworth & Co (Publishers) Ltd
London: 88 Kingsway, W.C.2B 6AB

AUSTRALIA

Butterworth & Co (Australia) Ltd
Sydney: 20 Loftus Street
Melbourne: 343 Little Collins Street
Brisbane: 240 Queen Street

CANADA

Butterworth & Co (Canada) Ltd
Toronto: 14 Curity Avenue, 374

NEW ZEALAND

Butterworth & Co (New Zealand) Ltd
Wellington: 49/51 Ballance Street
Auckland: 35 High Street

SOUTH AFRICA

Butterworth & Co (South Africa) Ltd
Durban: 33/35 Beach Grove

First published in 1971 by
Iliffe Books, an imprint of
the Butterworth Group

ISBN 0 592 02807 0

Printed in Hungary

Contents

Preface

The name 'thermistor' was coined as an acronym to designate a type of circuit element having a marked negative temperature coefficient of resistance over an extensive range of temperature: *therm*ally-sensitive res*istor*. These devices were developed in the 1940s as a result of materials research into the properties of metallic oxides mixed in non-stoichiometric proportions and to which, in some cases, trace 'impurity' atoms were added in very small proportions.

As a result of various 'recipes' being developed devices became commercially available from a number of firms. Their widespread use stemmed from a temperature coefficient of resistance of between -4 and -5% per degree Centigrade, which was considerably greater in magnitude than that of any element or metallic alloy. Further, since the mixed oxides had semiconducting, rather than metallic properties, a wide range of resistance values was available.

In the 1950s it was discovered that the ceramic barium titanate also had semiconducting properties, when a small proportion of either the barium or titanium atoms were replaced substitutionally by atoms of similar radii and of higher valency. Furthermore, over a limited range of temperature, above the Curie temperature, a marked positive temperature coefficient of resistance ($\approx 15\%$ per °C) arose.

It seemed logical to the author at the time that these materials, in device form, should also be classed as thermistors, but for some time 'thermistors' was in general reserved for the devices with negative temperature coefficient of resistance. Now, however, 'thermistor' is a generic term. Devices with negative temperature coefficient of resistance are designated N.T.C. and devices with positive temperature coefficient of resistance, P.T.C.

From the materials standpoint N.T.C. oxide thermistor development has not lead to any substantial increase in coefficient of resistance, although anomalous behaviour has been observed with certain mixes involving Vd_2O_5 as a constituent. This amounts to a negative temperature coefficient of resistance as high as 100% per °C over a limited range of temperature around 68 °C. Research is in hand to try to produce this behaviour at other temperatures by using different materials. Such research has been prompted by the rapid development of the P.T.C. barium titanate thermistor in which comparable proportional constituents of either lead or strontium titanate are involved, with controlled trace impurities introduced. Positive temperature coefficients of resistance as high as 60% per °C are now available over limited temperature ranges (a few tens of degrees) for tailored central temperatures between about −100 °C and +350 °C.

Other N.T.C. and P.T.C. materials have been developed with small (< 2% per °C) temperature coefficients of resistance and these have found some application in special cases.

Applications of thermistors are legion: they arise in all branches of science and engineering. Although measurement or indication of temperature is an obvious use, it has to be recognised that the resistance of a thermistor depends on both the ambient temperature and the internally dissipated power. The latter can either be due to applied terminal current and voltage or the absorption of electromagnetic radiation or energy from bombarding particles. Further, the overall temperature is a function of the thermal conductance of the thermistor. This is not a constant, but is a function of the ambient — whether it is liquid, gaseous or solid, and on the individual material involved.

The applications of both N.T.C. and P.T.C. thermistors are similar, although at present, if precision better than ±5% is required together with a high temperature coefficient of resistance, N.T.C. oxide types are favoured. These applications, which are treated in Chapters 5–9, are as follows: those dependent on (a) small-signal electrical parameters, (b) interaction with radiation, (c) variation of dissipation constant (or thermal conductance), (d) resistance-temperature characteristics, (e) circuit parameters. A final chapter deals with indirectly-heated (four-terminal) thermistors in which the thermistor material temperature is governed by the power dissipated in a separate metallic heater in thermal but not electrical contact with it.

The first four chapters of the book cover the material aspects and the representation of thermistors as electrical circuit elements and provide the necessary background for understanding the applications.

In such a small book it has not been possible to make acknowledgement to all the work that has been done in either developing or using thermistors, nor to go to any depth. In view of the absence (to the author's knowledge) of any comprehensive book on this extremely useful circuit component, it is hoped that the drawing together of even this limited amount of material will be of some use.

For one reason and another the typing of the manuscript has had a chequered history. My thanks are due to all who have been involved. Chronologically, these are, Susan Moyle, Jean Wooltorton, my wife, Janet Owens and, finally, Joan Radmore, who has had the difficult job of seeing through the final stages.

F. J. HYDE
Shrivenham 1969

The Publishers regret to record the death of the author, F. J. Hyde before his manuscript was set in type. This premature death at the age of 45 has removed from the Electronics scene a very able man. The present volume is the last of a long list of his publications and, like the others, bears the stamp of the man in the thoroughness of the treatment.

Mr. T. Williams, Principal Lecturer at the Royal Military College of Science, has completed the task of proof reading and his help is gratefully acknowledged.

CHAPTER 1

Introduction

1.1. TERMINOLOGY

Thermistor is an acronym for *therm*ally-sensitive-res*istor*. The electrical resistance of a thermistor is a function of its temperature, which depends on both the ambient temperature and atmosphere and on the internal power dissipation. Thermistors are made from semiconducting material and temperature coefficients between about −5% and +60% per K are normally encountered. The sign of the temperature coefficient distinguishes the basic types of thermistor.

Negative temperature coefficient (N.T.C.) thermistors are the more common and there is an extensive literature on the basic 'oxide' class. One of the key papers is that by Becker, Green and Pearson[1]; this was written at a time when the results of early workers[2–8] had been consolidated and there were already many applications. Oxide thermistors can be made with room temperature resistances ranging from a few ohms to megohms and the room temperature coefficient of resistance usually lies between −4% and −5% per K. Anomalous behaviour can, however, be obtained[9], with very large negative temperature coefficients over a short temperature range (≈10 K) at 68 °C. For normal N.T.C. thermistors the dependence of resistance on temperature approximately follows an exponential law. This is also true for single-crystal silicon carbide thermistors[10], which have application over the temperature range −100 to +300 °C.

Positive temperature coefficient (P.T.C.) thermistors are made from two kinds of material: compounds having the barium titanate structure[11–17] (one trade name for these is 'Posistors') and diamond-lattice type semiconductors such as silicon (a trade name here is 'Silistors'). The temperature range in which the barium titanate type of thermistor has a positive temperature coefficient of resistance is determined by composition and this coefficient can be very large (≈60% per K). Diamond-lattice type thermistors have a much smaller temperature coefficient of resistance (≈0·8% per K for silicon), but this is more uniform and applicable over a wider temperature range.

The majority of thermistors are two-terminal devices although three[18, 19] and four-terminal devices[1, 20] are made. The two- and three-terminal thermistors are referred to as directly-heated types, since the thermistor material itself absorbs power directly from the circuit in which it is connected. Four-terminal thermistors are termed indirectly-heated types. These have a separate heater in thermal, but not electrical, contact with the thermistor material.

1.2. FORM

Thermistors are made in a variety of shapes and sizes, although they are normally small. Various geometries have been developed to suit particular applications. These are referred to as beads (encapsulated in different ways), discs, washers, rods and thin films. Examples are shown in Fig. 1.1.

1.3. ELECTRICAL REPRESENTATION

To avoid complication at this stage, consideration will be given to the specification of the properties of standard two-terminal thermistors. This is done in terms of the following quantities:

R_a = the 'zero-power' d.c. resistance of the thermistor at the ambient temperature, T_a (K): this is the resistance measured when the applied power is so small that there is negligible self-heating.

R, $R(T)$ = the d.c. resistance under active conditions, T being the absolute temperature of the thermistor,

$P = VI$ = the power dissipated in the thermistor,

V = the voltage across the thermistor,

I = the current through the thermistor.

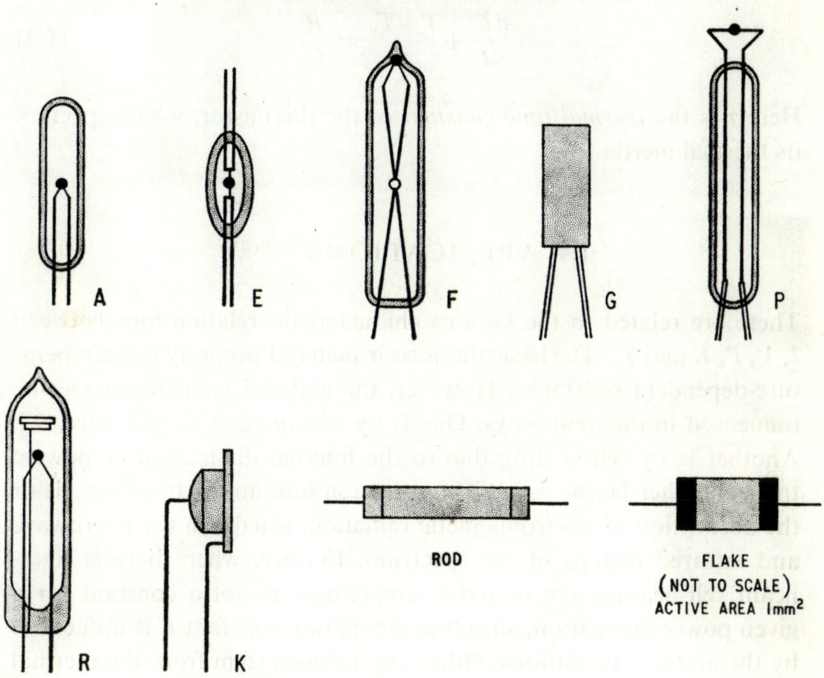

Fig. 1.1 Common thermistor shapes

In the steady state the equation relating the electrical to thermal power is

$$P = IV = K(T - T_a) \tag{1.1}$$

Here K is known as either the *thermal conductance* or *dissipation constant*, and is commonly quoted in mW/K. K is determined by both the construction of the thermistor and its operational environment. Provided that $T - T_a \ll T_a$, K is independent of T to a first order. Transient conditions are governed by the following equation;

$$C\frac{\mathrm{d}T}{\mathrm{d}t} + K(T - T_a) = P \tag{1.2}$$

where C is the *thermal capacity* of the thermistor. This equation may be written in the form

$$\frac{\mathrm{d}T}{\mathrm{d}t} + \frac{T - T_a}{\tau} = \frac{P}{C} \tag{1.3}$$

Here τ is the *thermal time constant* of the thermistor, which specifies its thermal inertia.

1.4. APPLICATIONS[16, 20, 21].

These are related to the various characteristic relationships between I, V, P, K and T_a. The basic thermistor material property is a temperature-dependent resistance. However, the material temperature can be influenced in different ways. One is by the ambient temperature, T_a. Another is by self-heating due to the internal dissipation of power: this can either be the result of connection into an electrical circuit or the absorption of electromagnetic radiation, usually in the microwave and infrared regions of the spectrum. Further, when there is significant self-heating, the material temperature is not a constant for a given power dissipation, since the dissipation constant K is influenced by the ambient conditions. Other applications stem from the thermal inertia, specified by the thermal time constant τ: these include both transient and small-signal conditions.

Typical V–I characteristics for N.T.C. and P.T.C. thermistors are shown in Fig. 1.2a and b, respectively. Here the parameter is T_a; the environment and hence K are invariant and there is no incident radiation. At low power levels, so that there is negligible self-heating, Ohm's Law is obeyed. As self-heating occurs so the N.T.C. thermistor resistance falls, whereas that of the P.T.C. thermistor rises. For most practical devices the power rating is such that the characteristic 'turnover' phenomenon occurs—i.e. there is a voltage maximum for N.T.C. and a current maximum for P.T.C. devices. In both cases, beyond turnover, the incremental resistance is negative. It may be noted here that for some P.T.C. thermistors a second turnover occurs

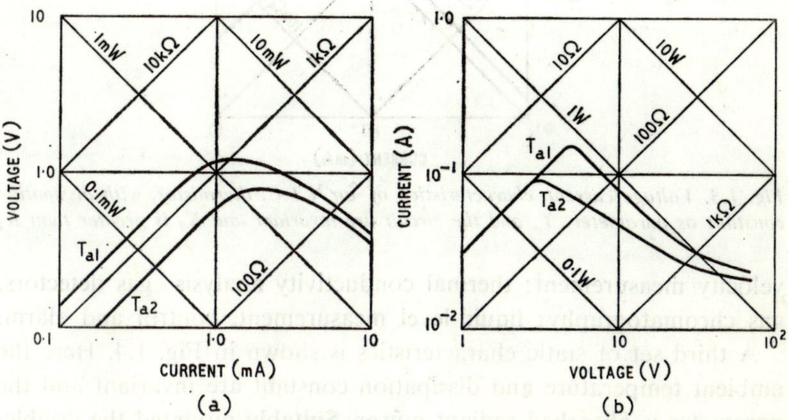

Fig. 1.2. Voltage/current characteristics of thermistors as a function of ambient temperature, T_a. The dissipation constant and circuit are invariant and $T_{a2} > T_{a1}$ (a) N.T.C. thermistor; (b) P.T.C. thermistor

due to contact non-linearity (see Section 3.3.1). Applications based on these characteristics are as follows: measurement and control of temperature, compensation of change of resistance with temperature in other circuit components; oscillator amplitude and/or frequency regulation, amplifier gain or level stabilisation and equalisation, voltage regulation (N.T.C. types), current regulation (P.T.C. types), speech volume limiting, expansion and compression; switching.

If the ambient temperature is kept constant then a further family of characteristics can be generated with the dissipation constant as the parameter, as shown in Fig. 1.3. For a given thermistor, *K* can be varied by connecting it to heat sinks having different thermal capacities, by changing the density of a gaseous environment, or the rate of flow of a surrounding fluid or gas. The following applications stem from these characteristics; vacuum manometers, flow meters, fluid

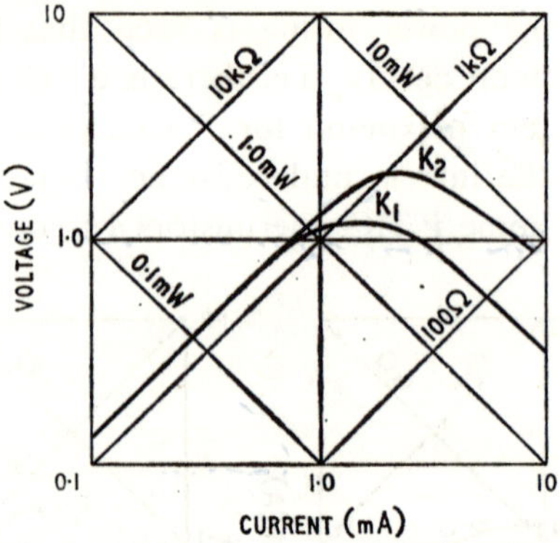

Fig. 1.3. Voltage/current characteristics of an N.T.C. thermistor, with dissipation constant as parameter. T_a and the circuit are invariant and K_2 is greater than K_1

velocity measurement; thermal conductivity analysis; gas detectors, gas chromatography; liquid-level measurement, control and alarm.

A third set of static characteristics is shown in Fig. 1.4. Here the ambient temperature and dissipation constant are invariant and the parameter is absorbed radiant power. Suitably mounted the double-ended type E thermistor (Fig. 1.1) can be arranged to absorb electromagnetic radiation in the microwave region and flake or thin-film thermistors behave similarly for infra-red radiation.

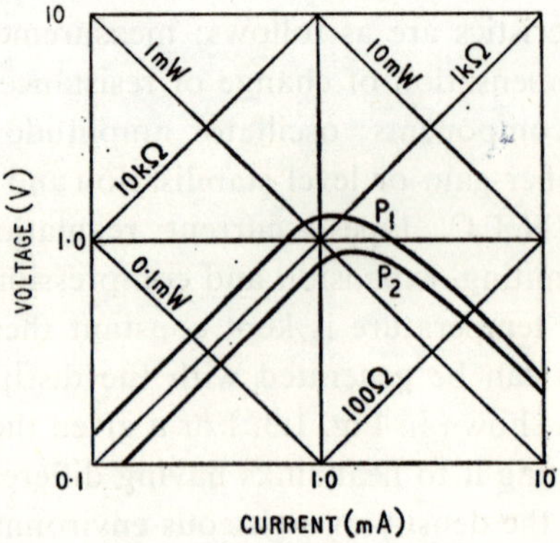

Fig. 1.4. Voltage/current, characteristics of an N.T.C. thermistor, with absorbed radiant power as parameter. K_1T_a and the circuit are invariant and P_2 is greater than P_1

Detailed analysis, based on Equation 1.3, shows that for a small-signal excitation, superimposed on a steady d.c. bias, the thermal inertia, which is specified by the thermal time constant τ, gives rise to inductive properties at frequencies of the order of the reciprocal of τ for N.T.C. thermistors and capacitive properties for P.T.C. thermistors. In consequence thermistors can be used in phase-advancing and phase-retarding networks and, if biased beyond turnover, so that the small-signal resistance is negative, in low frequency oscillators. At frequencies much greater than $1/\tau$ thermistors can be effectively used as variable radio-frequency resistors. For large-signal conditions the inertia represented by τ can be utilised in introducing delay, as in relay circuits and in surge suppression.

1.5. INDIRECTLY-HEATED THERMISTORS

A typical set of characteristics for an S.T.C. type 'B', N.T.C. thermistor, with ambient temperature and thermal conductance constant, is shown in Fig. 1.5. Heater current is the parameter. The same

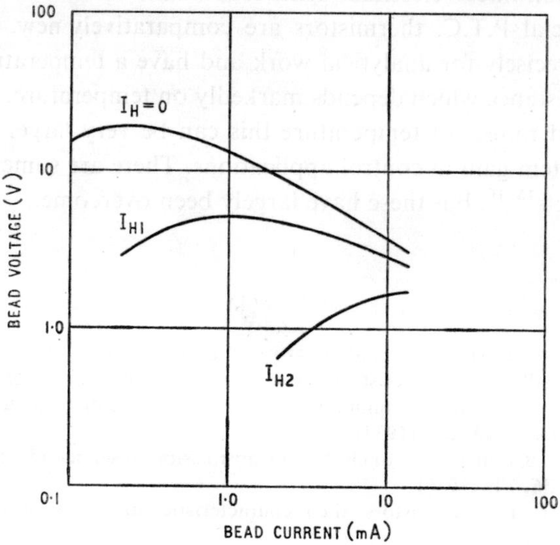

Fig. 1.5. Voltage/current characteristics of an indirectly-heated type B thermistor, with heater current as the parameter. K and T_a are invariant

characteristic voltage turnover as for directly-heated types is observed. Increased heater current is equivalent to an increase in ambient temperature in this set of curves. These thermistors, for which there is an additional means of controlling the thermistor resistance, other than by power dissipation in the bead itself, have useful applications in amplitude and gain control, timing circuits and the measurement of low frequency power. They can also be used in a.c./d.c. transfer instruments for power measurement at low frequencies.

1.6. N.T.C. OR P.T.C. THERMISTORS?

In subsequent chapters it will emerge that the N.T.C. thermistor, which has been available for some 25 years, is well established in circuit applications, particularly those where an analytical solution of the circuit problem is desirable. This is because the resistance-temperature dependence can be specified fairly closely by an exponential function. With N.T.C. thermistors there is also no problem with regard to non-linear electrical behaviour at the contacts.

Commercial P.T.C. thermistors are comparatively new, cannot be specified precisely for analytical work and have a temperature coefficient of resistance which depends markedly on temperature. However, over limited ranges of temperature this can be very large, with consequent system gain in control applications. There are some problems with contacts[15, 16], but these have largely been overcome.

REFERENCES

1. BECKER, J. A., GREEN, C. B., and PEARSON, G. L. 'Properties and uses of thermistors-thermally sensitive resistors' *Trans. Am. Inst. elect. Engrs* **65,** 711 (1946)
2. GUDDEN, B. 'Electrical conductivity of electronic semiconductors' *Ergebn. exakt Naturw.*, **13,** 223 (1934)
3. MEYER, W. 'Contribution to electrical transmission in semiconductor material' *Phys. Z.*, **36,** 749 (1935)
4. PEARSON, G. L. 'Thermistors, their characteristics and uses' *Bell Labs Rec.*, **19,** 106 (1940)
5. WEISE, E. 'Physical characteristics and technical applications of semiconductors' *Elektrotech. Z.* **59,** 1085 (1938)

6. SACHSE, H. 'Temperature-dependent resistances (hot conductors) and their technical applications' *Siemens-Z.* **19**, 214 (1939)
7. FAIRWEATHER, A. and INGHAM, J. 'Subsidence transients in circuits containing a nonlinear resistor, with reference to the problem of spark-quenching' *J. Inst. elect. Engrs.*, **88**, ptl, 330 (1941)
8. SILLARS, R. W. 'Materials and devices of falling resistance-temperature characteristic' *J. Scient. Instrum.*, **19**, 81 (1942)
9. FUTAKI, H. 'A new type semiconductor (critical temperature resistor)' *J. Appl. Phys. Japan*, **4**, 28 (1965)
10. BARTHA, H. D. and CARROLL, P. E. 'Unicrystalline silicon carbide thermistor' *I.R.E. Trans compon. Parts* **11**, 129 (1964)
11. SAUER, H. A. and FLASCHEN, S. S. 'Positive temperature coefficient of resistance thermistor materials for electronic applications' *Proc. electronic Compon. Symp.*, 41 (May, 1956)
12. SAUER, H. A., and FISHER, J. R. 'Processing of positive temperature coefficient thermistors' *J. Am. Ceram. Soc.*, **43**, 297 (1960)
13. SABURI, O., and WAKINO, K. 'Processing techniques and applications of positive temperature coefficient thermistors' *I.R.E. Trans compon. Parts* **10**, 53 (1963)
14. BRAUER, H., and FENNER, E. 'P.T.C. thermistors' *Siemens Rev.* **32**, 95 (1965)
15. WILSON, G. W., and SCHONBERG, S. 'An investigation of the constant current characteristics of materials having a positive temperature coefficient of resistance' *J. Appl. Phys.* **34**, 3055 (1963)
16. ANDRICH, E. 'Properties and applications of P.T.C. thermistors' *Electron. Applic. (Neth)*, **26**, 123 (1966)
17. IMAMURA, E., and TAMADA, M. 'Latest applications of MURATA POSISTOR. *Electron. Engng. Japan*, **12**, 60 (1967)
18. ASLAN, E. E. 'A single-bead broadband coaxial thermistor mount' *I.R.E. Trans. Instrumn.*, **14**, 102 (1965)
19. BERSIS, D. S., and KARYBAKES, C. A. 'A novel three-terminal thermistor; the varitherm' *J. Scient. Instrum.* **44**, 777 (1967)
20. SCARR, R. W. A., and SETTERINGTON, R. A. 'Thermistors, their theory manufacture and practice' *Proc. Instr. elect. Engrs*, **107B**, 395, 408, (1960)
21. SAPOFF, M., and OPPENHEIM, R. M. 'Theory and application of self-heated thermistors' *Proc. Inst. elect. electronic Engrs.*, **51**, 1292 (1963)

CHAPTER 2

Materials and Manufacture

2.1. INTRODUCTION

It was stated in the previous chapter that thermistors are made from 'semiconductors'. From an electrical standpoint a material is classified as a metal, semiconductor or insulator, depending on its resistivity, ϱ, although the boundaries between these classifications are somewhat arbitrary. At room temperature, if ϱ lies between 10^{-6} and 10^{-3} Ωcm a material is said to be a metal and if it is higher than 10^{12} Ωcm an insulator. Semiconducting materials have resistivities in between, although for practical purposes the room temperature range is limited to 10^{-3}–10^{7} Ωcm.

Semiconducting behaviour can arise for a variety of reasons. In N.T.C. oxide semiconductors it is governed by departures from stoichiometry (balance of elements as required by chemical formulae), in single-crystal N.T.C. and P.T.C. types by the presence of trace impurities in small atomic proportions (1 in 10^{6}–1 in 10^{10}), and by composition and impurities in perovskite-type P.T.C. devices. In this chapter this behaviour and an outline of manufacturing procedure will be given for the different types.

2.2. N.T.C. THERMISTORS

2.2.1. Oxide Semiconductors

N.T.C. thermistors are made from sintered mixtures of the oxides of certain metals. Such material is polycrystalline and the properties cannot be represented by a quantitative analysis based on order. Qualitatively, the presence of conducting carriers, either electrons or holes, can be explained in terms of either an excess or deficiency of metal atoms. Excess ionised metal atoms, over and above the requirement for stoichiometry (Frenkel defects) yield free electrons; i.e. these are not bound to oxygen atoms and are therefore available for conduction. Excess metal ions therefore give rise to *n*-type conductivity. If there are metal atoms missing from the lattice (Schottky defects) then adjacent oxygen atoms become negatively ionised, so giving rise to the creation of free holes and *p*-type conductivity.

2.2.2. Resistivity

Following the established theory for single-crystal semiconductors, a similar semiquantitative approach can be used to explain the macroscopic dependence of resistivity on temperature of oxide semiconductors[1]. Consider, for example, *n*-type material. If the density of valence electron sites in such a semiconductor is N, then at any absolute temperature T a fraction n/N of these will be free. The temperature-dependence of n is governed by an activation energy E_g, corresponding to an effective energy gap of about 0·5 V in practice, and in thermodynamic equilibrium

$$n^2 = (N-n)F \exp(-E_g/kT) \tag{2.1}$$

where k is Boltzmann's constant and F is a coefficient, which is slightly dependent on temperature. For $n \ll N$, as is the case in practice

$$n \simeq \sqrt{(FN)} \exp(-E_g/2kT) \tag{2.2}$$

In an *n*-type semiconductor the conductivity σ is given by $ne\mu$, where

μ is the electron mobility. μ depends on the particular scattering mechanisms involved, but can be represented over limited ranges of temperature by a power-dependence on temperature, as follows;

$$\mu \propto T^d \tag{2.3}$$

where d is a small negative number. Equations 2.2 and 2.3 can be combined to obtain the equation for resistivity originally given by Becker, Green and Pearson;[2]

$$\varrho = AT^{-c} \exp (B'/T) \tag{2.4}$$

Here c is a small positive or negative number or zero, B' usually lies between 1500 and 6000 K and A can have a wide range of values. Becker et al quote $\log_e A = 5{\cdot}563$, $c = 2{\cdot}73$ and $B' = 3100$ K for 'Material No. 1' and $\log_e A = 11{\cdot}514$, $c = 4{\cdot}83$ and $B' = 2064$ K for a form of 'Material No. 2'. Material No. 1 is composed of manganese and nickel oxides and Material No. 2 of oxides of manganese, nickel and cobalt. Fig. 2.1 shows the dependence of $\log_{10} \varrho$ on $1/T$ K^{-1} for these two materials.

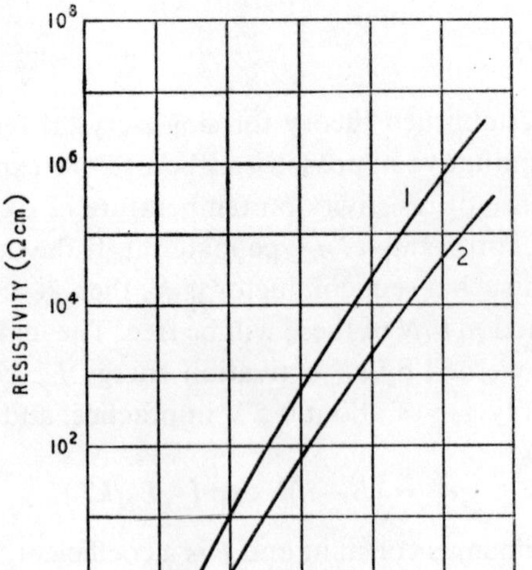

Fig. 2.1. Dependence of the resistivity of Bell Telephone Laboratories Materials 1 and 2 on the reciprocal of the absolute temperature (after Becker, Green and Pearson)

It will be noted that these log-linear plots approximate to straight lines, suggesting the following simpler form of dependence of resistivity on temperature for analytical purposes,

$$\varrho = \varrho_c \exp(B/T) \tag{2.5}$$

where B is a constant equal to 2·303 times the slope of the approximate straight lines. For Material No. 1 $B = 3\,920$ K, corresponding to an equivalent energy gap of 3 920/11 600 or 0·34 eV. Since $T = (P/K + T_a)$, from Equation 1.1, it follows that Equation 2.5 may be rewritten

$$\varrho = \varrho_c \exp B/(P/K + T_a) \tag{2.6}$$

ϱ and ϱ_c are related to corresponding terminal resistance values R and R_c by the same form-factor so that we may write, correspondingly;

$$R = R_c \exp(B/T) = R_c \exp\{B/(P/K + T_a)\} \tag{2.7}$$

For the 'zero-power' condition ($P \approx 0$), $R = R_a$, corresponding to the ambient temperature T_a and it follows that

$$R_a = R_c \exp(B/T_a) \tag{2.8}$$

and

$$R = R_a \exp\left\{B\left(\frac{1}{T} - \frac{1}{T_a}\right)\right\} \tag{2.9}$$

It may be noted that Bosson et al[3] found that the following empirical relationship gave a good fit to practical characteristics;

$$R = R_c \exp\big(B/(T + D)\big) \tag{2.10}$$

where D is a constant expressed in K.

2.2.2.1. TEMPERATURE COEFFICIENTS OF RESISTANCE

The temperature coefficient of resistance, α, may be defined as

$$\alpha = \frac{1}{R}\frac{dR}{dT} = \frac{d}{dT}\left(\log_e \frac{R}{R_a}\right) \tag{2.11}$$

Hence it follows from Equation 2.9 that

$$\alpha = -\frac{B}{T^2} \tag{2.12a}$$

For Bell Laboratories Material No. 1 and $T = 300$ K, $\alpha = -0{\cdot}044$. From the more accurate Equation 2.4

$$\alpha = -\frac{B'}{T^2} - \frac{c}{T} \tag{2.12b}$$

In Fig. 2.2 the dependence of α on resistivity at 20 °C is shown for S.T.C. material.

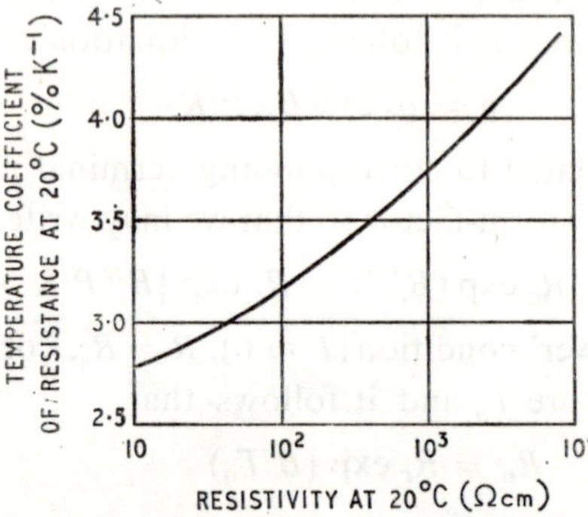

Fig. 2.2. Dependence of temperature coefficient of resistance on resistivity at 20 °C for oxide N.T.C. thermistor material (after Scarr and Setterington)

2.2.3. Materials

The earliest thermistors were made from uranium dioxide, from a magnesium oxide-titanium dioxide mixture and from copper oxide. In general these have been superseded because of manufacturing difficulties; e.g. it is necessary to sinter copper oxide at 1 000 °C under a pressure of 13 000 lb/in^2 of oxygen to obtain practical resistivity values.

The use of mixtures of metallic oxides, as described by Becker et al, was responsible for the present practice of manufacturing oxide semiconductors, in which sintering in air or other gaseous ambient is used and resistivity is basically controlled by varying the proportions of the constituent oxides. Oxides which have been used in various combinations and proportions are those of manganese, nickel, copper, cobalt, vanadium, chromium, titanium and tungsten. Other materials in commercial use are mixed ferrites and triferric-tetroxide containing traces of lithium oxide.

Basically there are now two distinct classes of material: that in which constituent oxides are mixed in comparable proportions and the other in which trace impurities are introduced as in single-crystal semiconductors.

2.2.4. Fabrication

It is possible to shape thermistors to given dimensions by one of the standard techniques such as (a) melting, cooling and solidifying; (b) evaporation; (c) heating compressed powders to a temperature at which they sinter into a strong, compact mass. For oxide thermistors the latter method is invariably used. The powder process makes it easy to mix two or more oxide components in various proportions and obtain a uniform and homogeneous solid. The resistivity of the material is determined mainly by the proportions of the ingredients and control is maintained by careful chemical analysis of the mixtures. Minor variations in resistivities can be brought about by control of the firing temperature, which lies between 1 000 and 1 350 °C.

As far as manufacture is concerned there are four basic shapes; bead, disc, rod and flake. In each case the basic oxides are intimately mixed and ground to a homogeneous fine powder in a ball-mill or similar device. The powder is then mixed with a suitable organic binder such as polyvinyl acetate to form a plastic body.

The consistency depends on the shape involved. For bead-type thermistors, two platinum-alloy wires of dia 0·001–0·004 in, are strung parallel to one another with a spacing of five to ten times the wire diameter. From 10 to 20 small masses of thermistor 'slurry' are applied at equal intervals and surface-tension draws these into beads as shown in Fig. 2.3. Size is carefully controlled using a projection-magnifier. They are allowed to dry and heated slightly until they have sufficient strength for the string to be handled. They are then passed through the sintering furnace. The beads shrink on to the wires and make intimate and permanent contact with them. The wires are cut at AA',

BB' and *CC'* in Fig. 2.3a to give the corresponding terminal connections shown in Fig. 2.3b. Electrical stability is achieved by protecting the bead from moisture, either by sealing in solid glass or in a glass bulb which is either evacuated or filled with a dried gas.

Platinum or one of its alloys is invariably used for the lead wires, since these have the required properties, namely; (a) a higher melting temperature than the sintering temperature of the thermistor; (b) a

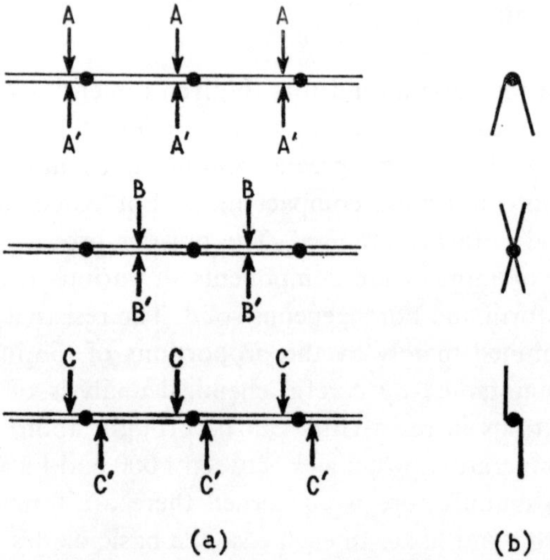

Fig. 2.3. Diagrammatic representation of the fabrication of different terminals for bead thermistors (after Scarr and Setterington)

coefficient of thermal expansion similar to that of the thermistor material; (c) resistance to oxidation in the case of thermistors sintered in oxidising atmospheres.

Rod thermistors are made by passing the slurry through a die, drying, cutting to length, heating to drive out the binder and then sintering; the temperature is raised gradually to avoid cracking and distortion. Contacts are made by coating the ends with silver, gold, or platinum paste, which is then cured before terminal contacts are made, either by pressing on wired end-caps or soldering on lead wires. Diameters are of the order of 0·08–0·64 cm and lengths 0·15–5 cm.

Disc and washer types are obtained by allowing the slurry to dry and then recrumbling to give grains which are formed into shape under high pressure in a die. Area metallic contacts are made, as for rod-type thermistors, and lead wires or a base-plate are then attached. Close tolerance for this geometry can be achieved as the resistance can be adjusted by removing small areas of metal from one face of the disc. Disc diameters vary from 0·15–5 cm and thicknesses from 0·08 to 0·64 cm.

Flake types start with a slurry of creamy consistency, which is spread as a film on a smooth glass surface and allowed to dry. The film is then removed, cut into flakes of the required size and shape and contacts are applied as before. Typical dimensions are: thickness 0·001–0·004 cm; length 0·1–1·0 cm; breadth 0·02–0·1 cm.

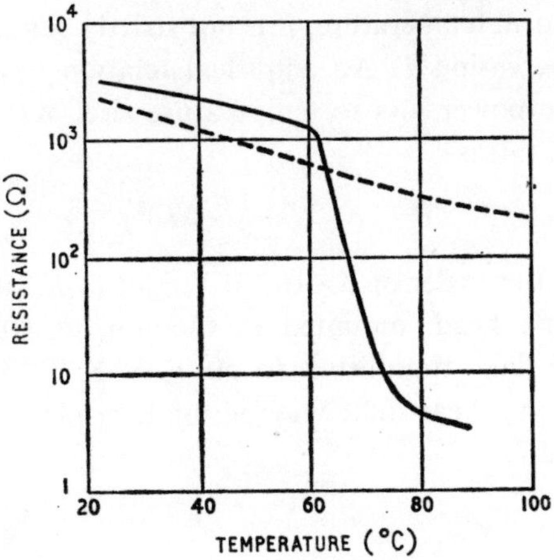

Fig. 2.4. Temperature dependence of resistance of a critical temperature resistor (full curve) and of a normal N.T.C. thermistor (broken curve). (After Futaki)

Anomalous, but reproducible behaviour has been achieved by sintering a mixture of vanadium pentoxide and an acidic oxide (B, Si or P oxide) or a basic oxide (Mg, Ca, Sr or Ba oxide), in a suitable reducing atmosphere,[4, 5]. The temperature coefficient of resistance is always negative, but in the immediate vicinity of 68 °C becomes extremely large. This is shown in Fig. 2.4, in which a comparison is made of the resistance-temperature characteristic of such a 'critical temperature resistor' (full curve) and a normal oxide thermistor (broken curve).

2.2.5. Dissipation Constant, *K*

This was introduced in Equation 1.1 and may be formally defined as the power required for a unit temperature rise. It is alternatively known as the thermal conductance. *K* depends on the environment of the thermistor and on its mounting as well as on its ratio of surface to volume and the thermal conductivity of the lead wires and envelope. The conditions under which it is measured should therefore be specified. Since cooling is not strictly Newtonian, nor is the thermistor at a uniform temperature, *K* is not strictly constant: it increases slightly with increasing *T*. An empirical relation has been given by Smith[6] to relate power loss to temperature rise, $\Delta T = T - T_a$; when ΔT is not small, this is

$$P = K(\Delta T + k''(\Delta T)^6) \tag{2.13}$$

where k'' is of the order of 3×10^{-12}. Values of *K* range from about 10 μW K^{-1} for a bead, mounted in vacuum, or 50 μW K^{-1} for a vacuum-backed flake thermistor to 50 mW K^{-1} for large rods or discs mounted on a heat-sink. Methods of determining *K* are described in Section 3.2.7.

2.2.6. Thermal Time Constant, τ[1, 7–10]

This was introduced in Equation 1.3 as the quotient of the thermal capacity *C* and dissipation constant *K*. The thermal capacity is the product of the mass and specific heat and is therefore governed by the volume and the density. A large mass and small dissipation constant means a large τ and vice versa. The cooling curve for a massive rod-type thermistor in air is given in Fig. 2.5. This was obtained by switching over at $t = 0$ to a constant current supply of small magnitude, so

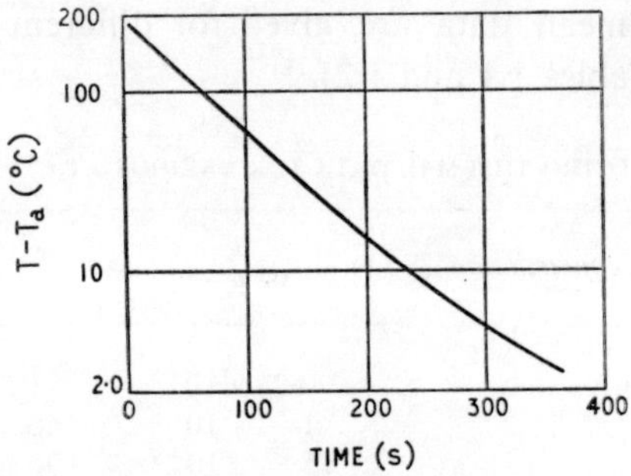

Fig. 2.5. Cooling characteristic of a massive N.T.C. thermistor presented as log (excess temperature above ambient) versus time. (After Becker, Green and Pearson)

that I^2R was at all times negligible and the voltage across the thermistor was determined as a function of time; thence R and T were computed. In Table 2.1 deduced data for K, τ and C are given as a function of T.

Table 2.1. ELECTRO-THERMAL DATA FOR ROD-TYPE THERMISTOR[1]

T	K	τ	C
°C	W °C^{-1}	s	J °C^{-1}
44	$3{\cdot}7\times10^{-3}$	76	0·28
64	$3{\cdot}7\times10^{-3}$	74	0·27
84	$3{\cdot}8\times10^{-3}$	71	0·27
104	$3{\cdot}7\times10^{-3}$	69	0·26
124	$3{\cdot}8\times10^{-3}$	68	0·26
144	$3{\cdot}8\times10^{-3}$	67	0·26
164	$3{\cdot}9\times10^{-3}$	67	0·26
184	$4{\cdot}1\times10^{-3}$	66	0·27
204	$4{\cdot}2\times10^{-3}$	66	0·28

The relative insensitiveness of C to T is evident, as is the decrease in τ with increase in T, due to the expected rise of K with T. The specific heat of oxide-type N.T.C. thermistor material is approximately 0·73 J g^{-1} K^{-1} and its density about 4·5 g cm^{-3}.

In Table 2.2 mean data are given for different types of S.T.C. thermistor (see Tables 1.1 and 1.2).[11]

Table 2.2. ELECTRO-THERMAL DATA FOR VARIOUS S.T.C. THERMISTORS

Thermistor construction	K W K^{-1}	τ s	C J K^{-1}
Bead in gas-filled bulb	$0{\cdot}28\times10^{-3}$	1·5	$0{\cdot}42\times10^{-3}$
Rod	1×10^{-3}	140	2·52
Disc	15×10^{-3}	90	1·35
Disc on mounting plate attached to heat sink	100×10^{-3}	33	3·3
Bead in glass	2×10^{-3}	24	0·048

Departures from a near-exponential dependence are observed if the thermistor is not surrounded by air but by a solid material such as glass or resin. There is an initial rapid decrease, which is followed by an exponential one. This phenomenon does not appear to have been investigated in detail.

2.2.7. Ageing[1, 11, 12]

To obtain stable resistance/temperature characteristics thermistors are aged. This is a lengthy process, especially for thermistors used for precise thermal measurements. For example, a thermistor may be measured at 20 °C, stored for a week and rejected if it shows a resistance change of more than 1% during this period. Long-term stability is largely governed by the ageing process and can be affected by

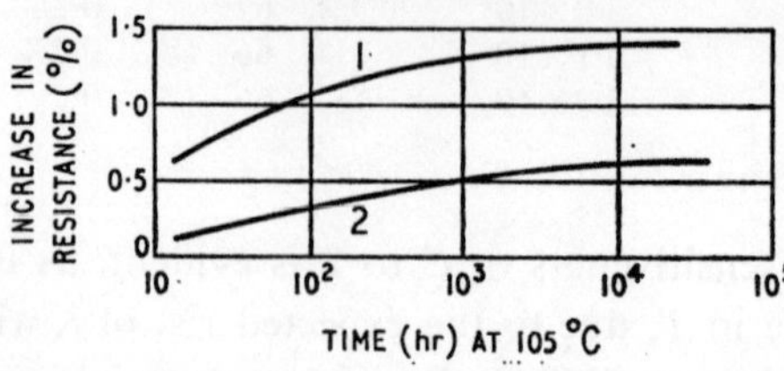

Fig. 2.6. Ageing characteristics of Bell Telephones materials Nos 1 and 2. (After Becker, Green and Pearson)

thermal shock or large temperature changes. Thermistors should be cycled over a temperature range in excess of the working range prior to calibration. There are no general conclusions with regard to long-term stability as many varied factors are involved. Examples of change in resistances in thermistors aged in an oven at 105 °C are given in Fig. 2.6 for Bell Laboratories thermistors made from Materials 1 and 2.

2.2.8. Unicrystalline Silicon Carbide Thermistors[13]

Controlled resistivity of silicon carbide crystals can be achieved using boron as a dopant; crystals used for thermistor manufacture are selected on the basis of mechanical perfection and tungsten lead wires are welded on either side, using a current pulse. The boron doping level is chosen to give a workable room-temperature resistivity. The resistivity is a function of temperature, primarily because of the increase of intrinsic carrier concentration with temperature, which overrides the corresponding increase in scattering and hence the decrease in the mobility of the free holes. The approximately exponential dependence of resistivity on $1/T$ observed experimentally is shown in Fig. 2.7. The room temperature coefficient of resistance is comparatively low ($-1{\cdot}9\%$ per K), but the operating temperature range is from -100

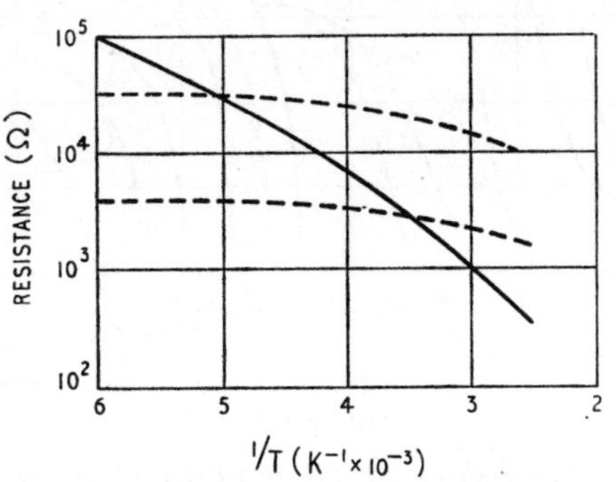

Fig. 2.7. Dependence of the resistance of optimum-doped SiC thermistors on (temperature)$^{-1}$. (full curve) This curve represents the superposition of five sample characteristics. The broken curves show the effect of lower doping levels. (After Batha and Carroll)

to +300 °C, with good retraceability of characteristics after temperature cycling. Typical devices made by the Carborundum Company had a zero power resistance of 2 600 Ω, a dissipation constant of 60 μW K^{-1} and a thermal time constant of ≈0·25 s.

2.3. P.T.C. THERMISTORS

2.3.1. Barium titanate types

The discovery that doped barium titanate, which is inherently a non-conducting ceramic, exhibited a marked positive temperature coefficient of resistance over a relatively short range of temperature is due to Haayman et al[14]. It has beem exploited by other workers, some of whom are listed in references 11–16 in Chapter 1. A useful range of thermistors is now available, in which doped material of the general complex, barium-strontium-lead titanate is used. Andrich[15] has shown

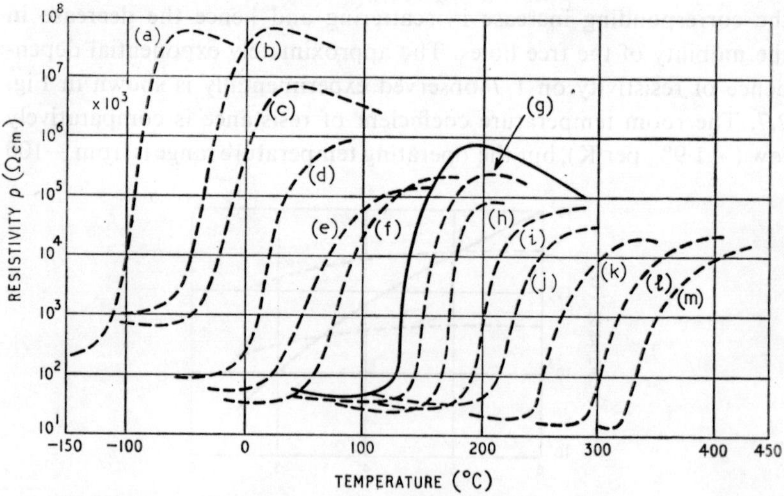

Fig. 2.8. Temperature-dependence of the resistivity of material of the barium-strontium-lead titanate complex. All materials contain $BaTiO_3$*. Samples on the left of the solid curve contain* $SrTiO_3$ *in the following percentages: (a) 70, (b) 60, (c) 50, (d) 40, (e) 30, (f) 20. Samples on the right contain* $PbTiO_3$ *in the following percentages: (g) 5, (h) 10, (i) 15, (j) 20, (k) 30, (l) 40, (m) 50. (After Andrich)*

resistivity versus temperature curves for this system in which the relative proportion of strontium to lead titanate is varied from (70:0) to (0:50). These curves are reproduced in Fig. 2.8.

Brauer and Fenner[16] have given a semiquantitative account of the resistance anomaly in relation to barium titanate. Barium titanate is a ferro-electric material and has a high dielectric constant, ε, which reaches a maximum in the vicinity of the Curie point (115 °C), for low field intensities and falls thereafter, in conformity with the Curie-Weiss law,

$$\varepsilon(T) = \frac{C}{T - T_c} \tag{2.14}$$

where $C = 1{\cdot}2\times10^5$ is the Curie constant, T_c the Curie point and T the operating temperature. This is shown as the full curve in Fig. 2.9.

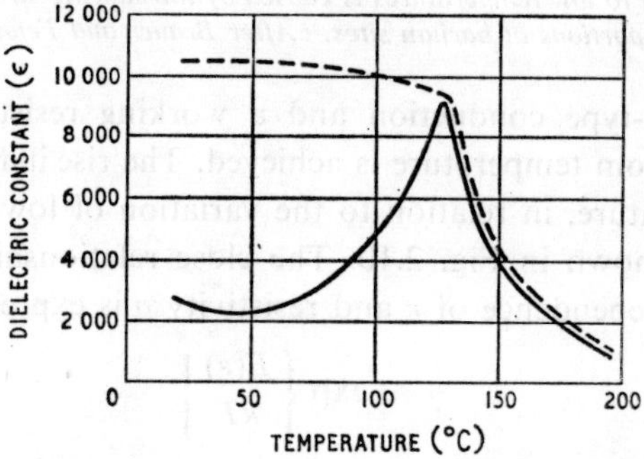

Fig. 2.9. Dependence of the dielectric constant of $BaTiO_3$ *on temperature. The continuous curve is for low-field conditions (≈10 V/cm) and the broken curve for high-field conditions (≈3 kV/cm). (After Brauer and Fenner)*

The dashed curve arises for high-field conditions. The semiconducting behaviour is brought about by replacing barium with atoms of elements having similar ionic radii to Ba^{2+} and valencies greater than two or those having ionic radii similar to Ti^{4+} and valencies greater than four. Brauer and Fenner used Sb^{3+} in small concentrations (< 0·3 atom %), thereby producing the valence change $Ti^{4+} \rightarrow Ti^{3+}$. This means that in the vicinity of titanium ions an equivalent number of excess electrons are produced and these are free to move within the lattice. There

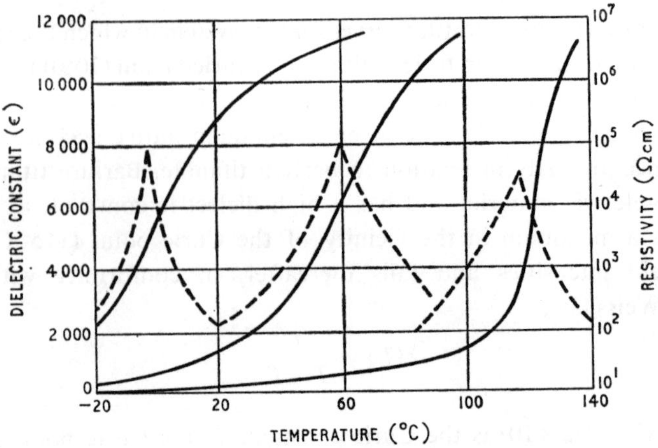

Fig. 2.10. Dependence of resistivity on temperature of $BaTiO_3$ *thermistors (continuous curves) compared with corresponding variation of dielectric constant (broken curves). The shift to low temperatures is caused by introducing strontium in different proportions at barium sites. (After Brauer and Fenner)*

is therefore *n*-type conduction and a working resistivity of about 30 Ω cm at room temperature is achieved. The rise in resistivity at the Curie temperature, in relation to the variation of low field dielectric constant, is shown in Fig. 2.10. The close relationship between the temperature-dependence of ε and resistivity ϱ is expressed by

$$\varrho \approx \exp\left\{\frac{E(\varepsilon)}{kT}\right\} \tag{2.15}$$

where $E(\varepsilon)$ is the activation energy of surface states which are assumed to arise at the barrier layers existing between individual crystalline

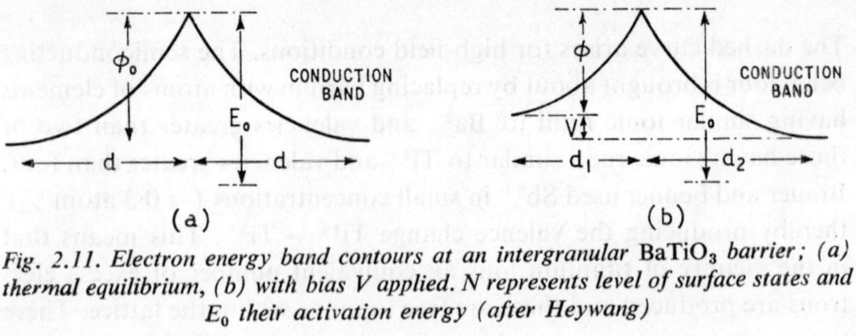

Fig. 2.11. Electron energy band contours at an intergranular $BaTiO_3$ *barrier; (a) thermal equilibrium, (b) with bias V applied. N represents level of surface states and* E_0 *their activation energy (after Heywang)*

grains. The electron energy-band diagram of such a barrier[17] is shown in Fig. 2.11. The potential barrier, ϕ, which has to be overcome by conduction electrons in thermal equilibrium is given by

$$\phi_0 = \frac{e^2 N_D d^2}{2\varepsilon\varepsilon_0} \qquad (2.16a)$$

and when a potential V is applied by

$$\phi = \phi_0\left(1 - \frac{V}{4\phi_0}\right) \qquad (2.16b)$$

Here e is the electronic charge and d the half-width of the depletion layer in thermal equilibrium. For large values of ε such as arise below the Curie point, the potential barriers ϕ_0 and ϕ and hence the resistivity are small. These large values of ε follow the dashed curve of Fig. 2.9 since high-field intensities arise in the depletion layers. If the temperature exceeds the Curie point ε drops sharply and d and ϕ increase rapidly. The result is the observed increase in resistivity. This is clearly voltage-dependent since the application of an external voltage modifies the height of the potential barrier. It is independent of polarity since the barrier is symmetrical.

2.3.1.1. FABRICATION

A detailed account of fabrication techniques and of the effectiveness and optimum concentrations of different dopants is given by Saburi and Wakino[18]. They state that, 'Although the starting formulation can be a combination of $BaCO_3$, $SrCO_3$, TiO_2, SiO_2 and La_2O_3, a superior formulation is barium titanate and strontium titanate prepared from barium- and strontium-titanyl oxalate, tin oxide and lanthanum oxalate. The latter formulation is preferable because of the higher purity that can be attained and the superior stoichiometry of the material'.

Production is similar to that for N.T.C. oxide types, involving grinding of the constituents in a ball-mill with water present, drying of the slurry, pulverising and then calcining for an hour at 1 100–1 200 °C. During the calcining the materials usually form white and brittle lumps, which are then crushed in a ball-mill with water present, dried and repulverised into powder.

This powder is pressed into plates, discs or bars hydraulically and the pressed bodies are then fired at a temperature between 1 300 and 1 400 °C for 0·5–2 hr: both the firing temperature and the soaking time critically affect the end-product.

It has been found that near-ohmic contacts to the bodies can be formed using the electrodeless nickel deposition technique developed by Turner et al.

2.3.1.2. D.C. RESISTIVITY

The range of resistivity-temperature curves obtained by Saburi and Wakino,[18] on a reproducible basis, is shown in Fig. 2.12. Each charac-

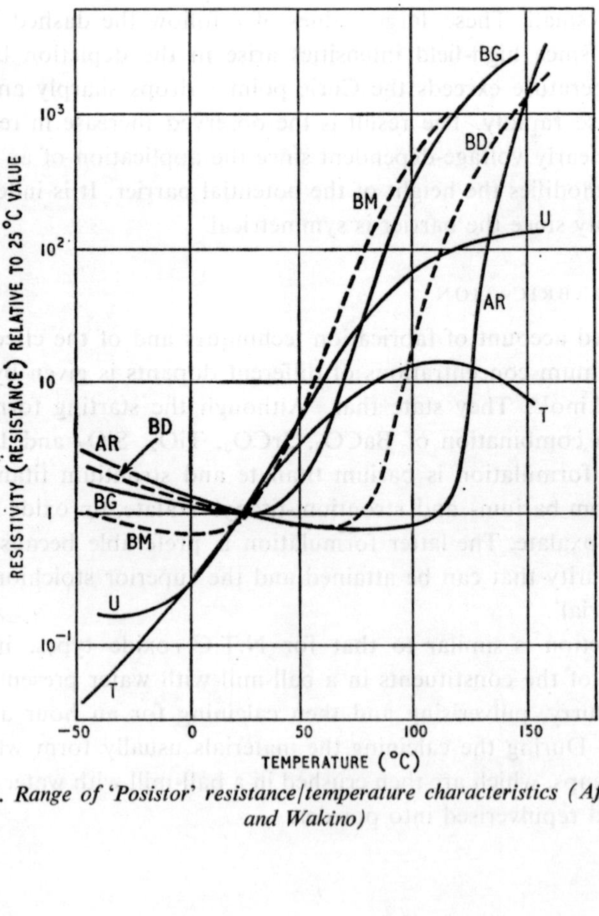

Fig. 2.12. Range of 'Posistor' resistance/temperature characteristics (After Saburi and Wakino)

teristic can be regarded as part of the prototype characteristic shown in Fig. 2.13, which has a negative resistivity-temperature characteristic up to a temperature T_p, a positive temperature characteristic between T_p and T_n, with a maximum temperature coefficient at T_m and a negative temperature coefficient above T_n. Between T_p and T_n Saburi

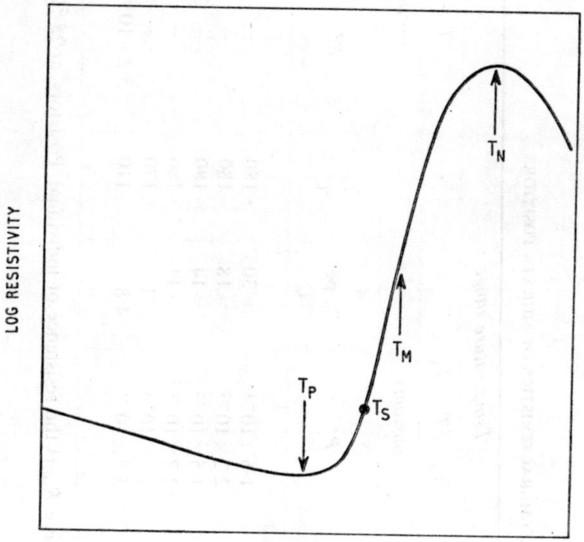

Fig. 2.13. Prototype resistivity/temperature characteristic of 'Posistor' material

and Wakino assume an exponential dependence of resistance on temperature, complementary to the simplified form of Equation 2.7 which has been widely used for N.T.C. thermistors; i.e.

$$R = R_c \exp AT \tag{2.17}$$

so that $\alpha = A$, in this case. In the negative resistance ranges they utilise Equation 2.7. Values of A, B and R_c for the different characteristics of Fig. 2.12 are given in Table 2.3.

Manufacturers of these thermistors usually specify a switching temperature T_s at which the resistance rises to twice the cold value.

Table 2.3. CHARACTERISTICS OF MURATA POSISTORS

Type of characteristic	Temperature range								
	$T < T_p$			$T_p < T < T_n$			$T > T_n$		
	constants			constants			constants		
	T_p °C	P*	B K	T_m °C	P*	A % per °C	T_n °C	P*	B K
AR	110	$9 \cdot 1 \times 10^{-2}$	680	130	$1 \cdot 6 \times 10^{-16}$	>30	>180	—	—
BD	80	$9 \cdot 1 \times 10^{-2}$	680	110	$2 \cdot 2 \times 10^{-7}$	>18	>180	—	—
BG	50	$7 \cdot 1 \times 10^{-2}$	750	75	$1 \cdot 5 \times 10^{-3}$	>14	>180	—	—
BM	15	$1 \cdot 0 \times 10^{-1}$	530	65	$2 \cdot 7 \times 10^{-2}$	>11	>180	—	—
U	—20	$7 \cdot 1 \times 10^{-3}$	750	55	$1 \cdot 7 \times 10^{-1}$	7·4	>170	—	—
T	<—50	—	—	50	$3 \cdot 2 \times 10^{-1}$	4·8	110	$5 \cdot 2 \times 10^{-3}$	$3 \cdot 1 \times 10^{-3}$

* The values of R_c are given by $R_c = PR_{p0}$, where R_{p0} is the resistance of individual 'Posistors' at 25 °C.

2.3.1.3. A. C. RESISTIVITY

Barium titanate type P.T.C. thermistors differ from N.T.C. and silicon P.T.C. types, because of the intergranular barriers that appear to exist. It has been demonstrated that these control the overall resistivity above the Curie temperature. Since they are space-charge regions they also contribute a capacitive effect. In consequence the local barriers behave as imperfect dielectrics, and in consequence the impedance they present is frequency-dependent. Andrich[15] states that, 'It has been experimentally demonstrated that the ratio R_{max}/R_{min} varies inversely with frequency, and at frequencies greater than 1 MHz,

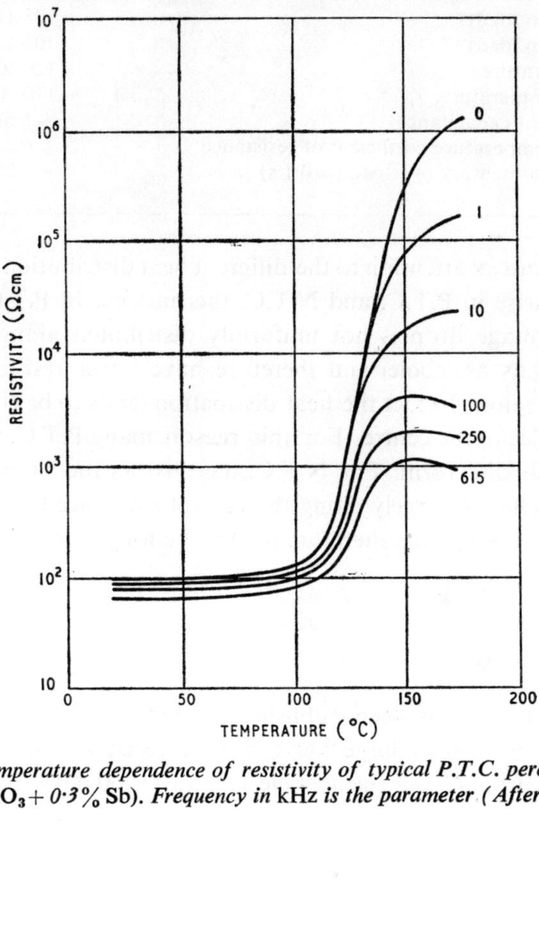

Fig. 2.14. Temperature dependence of resistivity of typical P.T.C. perovskite material ($BaTiO_3$ + *0·3%* Sb). *Frequency in* kHz *is the parameter (After Andrich)*

there is no longer any positive temperature coefficient at all.' The trend can be seen in the curves of Fig. 2.14. At the lowest frequencies the frequency-dependent effects due to thermal inertia, which are the only significant ones for N.T.C. thermistors, will also be involved.

Typical properties of barium titanate thermistors have been given by Andrich[15] and are reproduced in Table 2.4.

Table 2.4. GENERAL DATA FOR BARIUM TITANATE THERMISTORS

Specific heat	0·12 cal/g K
Specific resistivity at 25 °C	20–50 Ωcm
Specific weight	5·6 g/cm³
Dissipation constant in still air (samples with volume ≈0·5 cm³)	10–15 mW/K
R_{max}/R_{min} (no load)	10^4–10^5
R_{max}/R_{min} (on load)	10^2–10^3
Curie temperature	120 °C
Switching temperature, T_s	110 °C
Inherent shunt capacitance	~1 nF
Maximum temperature coefficient of resistance	~60%/K
Maximum momentary overload (≈0·1 s)	~1 kW

He also draws attention to the different heat distributions which are bound to arise in P.T.C. and N.T.C. thermistors. In P.T.C. thermistors the voltage drop is not uniformly distributed along its length since the ends are cooler and therefore have lower resistance in the switching region. In fact the heat dissipation tends to be concentrated in a thin slice in the centre. For this reason many P.T.C. thermistors are made in disc form. For N.T.C. thermistors the situation is that the current flow is largely along the central axis since the periphery is cooler. Rod thermistors therefore tend to be long and slender rather than fat.

2.3.1.4. AGEING

Not a great deal has been published on long term stability. Saburi and Wakino[18] report large variations of resistance initially but a settling to within 6% of the average value after 60 days.

2.3.2. Silicon Thermistors

The conductivity σ of a specimen of silicon, a tetravalent element, is given by

$$\sigma = ne\mu_e + pe\mu_h \tag{2.18}$$

where n and p are the electron and hole densities, respectively, μ_e and μ_h are the corresponding mobilities and e is the magnitude of the electronic charge. In general, by controlled doping, extrinsic material is made, so that either electron or hole conductivity is dominant, giving rise to n-or p-type material, respectively. Hence

$$\sigma_n = ne\mu_e \tag{2.19a}$$

and

$$\sigma_p = pe\mu_h \tag{2.19b}$$

The dependence of conductivity on temperature for both n- and p-type material is shown in Fig. 2.15. The parameter in part (a) is the concentration, N_d, of donor (pentavalent) atoms, each of which provides a free (conduction) electron at all temperatures above about 20 K, and in part (b) the concentration, N_a, of acceptor (trivalent) atoms, which correspondingly produce free holes.

The dependence of σ on temperature is governed by the combined variations of charge carrier density and mobility with temperature. In extrinsic material the charge carrier density is determined by the doping level. There is in all cases an increase in both p and n as the temperature is raised due to the thermal generation of hole-electron pairs. Over the temperature range shown, μ_e and μ_h decrease steadily with increasing temperature due to the increasing effectiveness of lattice scattering. At the lower temperatures the variation of mobility is dominant and there is a positive temperature coefficient of resistance. This is greater in magnitude the smaller is the concentration of dopant atoms, but for these conditions is limited in its manifestation, since as the temperature is increased the increase in densities n and p eventually overide the fall in the mobilities. Because the minimum conductivity occurs at a temperature some 20–30 °C higher for n-type than for p-type material, the former is generally used for P.T.C. silicon thermistors.

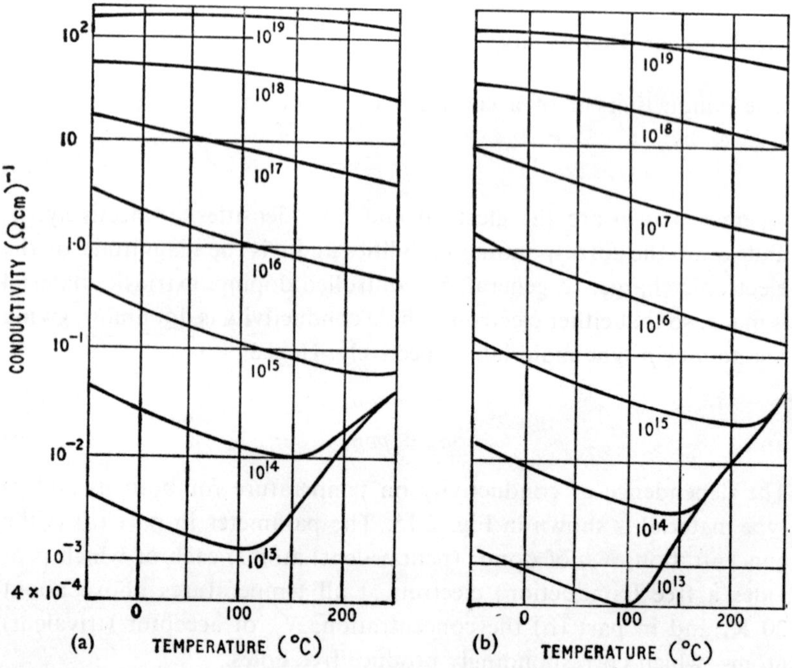

Fig. 2.15. Conductivity of silicon as a function of temperature: dopant concentration is the parameter; (a) n-type, (b) p-type

For 'Silistors' made by Standard Telephones and Cables Ltd., the following empirical relationship between resistivity ϱ to that, ϱ_{25}, at 25° C, arises;

$$\varrho = \varrho_{25}\left(\frac{273+T}{298}\right)^{2\cdot 3} \tag{2.20}$$

where T is in °C. The relation between 'cold' room temperature resistance R_a (internal power dissipation zero) and resistance R with internal power dissipation P (watts) is

$$R = R_a(1+0\cdot 7P)$$

A typical value for K is 11 mW/K and resistances, R_a between 10 and 680 Ω are available.

REFERENCES

1. GOUDET, G. and MEULEAU, C. *Semiconductors, their theory and practice*, Macdonald and Evans (1957)
2. BECKER, J. A., GREEN, V. B., and PEARSON, G. L. 'Properties and uses of thermistors-thermally sensitive resistors', *Trans. Am. Inst. elect. Engrs.*, **65**, 711 (1946)
3. BOSSON, G., GUTMAN, F., and SIMMONS, L. M. 'A relationship between resistance and temperature of thermistors', *J. Appl. Phys.* **21**, 1267 (1950)
4. FUTAKI, H. 'A new type semiconductor (Critical temperature resistor)', *J. Appl. Phys. Japan*, **4**, 28 (1965)
5. MORIN, F. J. 'Oxides which show metal-to-insulator transition at the Néel temperature; *Phys. Rev. Lett.* **3**, 34 (1959)
6. SMITH, O. J. M. 'Thermistors: Part I; Static characteristics and Part II; Dynamic characteristics', *Rev. scient. Instrum.*, **21**, 344, 351 (1950)
7. EKELÖF, S., and KIHLBERG, G. 'Theory of the thermistor as an electric circuit element', *Chalmers tek. Högsk. Handl.* (1954) *Avd Elektroteknik* 40 Nr. 142
8. CANDY, C. J. N. 'The specification of the properties of the thermistor as a circuit element in very low-frequency systems', *Proc. Instn. elect. Engrs.*, **103B**, 398 (1954)
9. HYDE, F. J. 'The impedance of a thermistor at low frequencies', *J. Electronic.*, **1**, 303 (1955)
10. EKELÖF, S., BJÖRK, N. and DAVIDSON, R. 'Large-signal behaviour of directly-heated thermistors, *Chalmers tek. Högsk. Handl.*, (1957) *Avd Elektroteknik*, **56**, Nr. 185
11. SCARR, R. W. A., and SETTERINGTON, R. A. 'Thermistors, their theory, manufacture and application', *Proc. Instn. elect. Engrs*, **107B**, 395, 408 (1960)
12. BECK, A. 'The stability of thermistors; J. Scient. Instrum. **33**, 16 (1956)
13. BATHA, H. D. and CARROLL, P. E. 'Unicrystalline silicon carbide thermistor', *I.R.E. Trans compon. Parts*, **11**, 129 (1964)
14. HAAYMAN, P. W., DAM, R. W., and KLASSENS, H. A. 'Method of producing semiconducting material', *German Patent No. 929350* June 23 1955
15. ANDRICH, E. 'Properties and applications of P. T. C. thermistors', *Electron. Applic.* **26**, 123 (1966)
16. BRAUER, H., and FENNER, E. 'P.T.C. thermistors', *Siemens Rev.* **32**, 95 (1965)
17. HEYWANG, W. 'Bariumtitanat als Sperrschichthalbleiter', *Solid St. Electron.*, **3**, 51 (1961)
18. SABURI, O. and WAKINO, K. 'Processing techniques and applications of positive temperature coefficient thermistors', *I.R.E. Trans. compon. Parts* **CP–10**, 53 (1963)

CHAPTER 3

D.C. Characteristics

3.1. INTRODUCTION

In the subsequent discussion of d.c. characteristics, Newtonian cooling will be assumed, so that the steady-state relation between electrical power, P, dissipated in the thermistor material as heat loss per second is

$$P = VI = K(T-T_a) \tag{3.1}$$

with K independent of T for $T-T_a \ll T_a$. It is implicit that all parts of the thermistor are assumed to be at the same temperature, T. For large values of $T-T_a$ it is necessary to use the more complicated expression due to Smith (see Equation 2.13). It will be assumed that non-linear contact phenomena, such as arise for some barium-titanate based thermistors, are ignored. Indeed, a detailed analytical treatment of the characteristics of this kind of thermistor is not possible, for applications in which precision is required.

3.2. N.T.C. THERMISTORS

The experimental dependence of the resistance of normal oxide-semiconductor thermistors on temperature is presented in Fig. 2.1.

Various equations relating R and T, due to Becker et al[1], have also been given. In this section the properties of this type of thermistor will be considered in more detail.

3.2.1. Specific Definitions[2]

3.2.1.1. POWER SENSITIVITY

From Equations 2.9 and 3.1 we can write

$$-\log_e\left(\frac{R}{R_a}\right) = \frac{B(T-T_a)}{TT_a} = \frac{BP}{T_a(P+KT_a)} \tag{3.2}$$

The power sensitivity, which specifies the change of resistance with power dissipation, is defined as

$$p = -\frac{1}{R}\frac{\mathrm{d}R}{\mathrm{d}P} = -\frac{\mathrm{d}}{\mathrm{d}P}\left\{\log_e\left(\frac{R}{R_a}\right)\right\} \tag{3.3}$$

so that on substitution for $\log_e(R/R_a)$ from Equation 3.2 we obtain

$$p = \frac{B}{KT_a^2}\,\frac{1}{(1+P/KT_a)^2} = \frac{p_0}{(1+P/KT_a)^2} \tag{3.4}$$

Here $p_0 = B/KT_a^2$ is the power coefficient of resistance when $P/KT_a \ll 1$ For high sensitivity B/K should clearly be as large as possible. It is evident from Equation 3.4 that the sensitivity falls to 25% of its maximum value when $P = KT_a$; i.e. at $T = 2T_a$.

3.2.1.2. TEMPERATURE SENSITIVITY

The temperature coefficient of resistance is defined as follows:

$$\alpha = \frac{1}{R}\frac{\mathrm{d}R}{\mathrm{d}T} = \frac{\mathrm{d}}{\mathrm{d}T}\left\{\log_e\left(\frac{R}{R_a}\right)\right\} \tag{3.5}$$

Hence, differentiating Equation 3.2 with respect to T, we find that

$$\alpha = -\frac{B}{(P/K+T_a)^2} \tag{3.6}$$

which reduces to $-B/T_a^2$ when $P/K \ll T_a$.

3.2.1.3. HALF-TEMPERATURE

It is useful to know the temperature change, ΔT, which will halve or double the resistance in relation to its value at an arbitrary reference temperature T_a. From Equation 3.2

$$\frac{B}{T_a \pm \Delta T} - \frac{B}{T_a} = \pm \log_e 2 \tag{3.7}$$

so that

$$\Delta T = \frac{\log_e 2}{-\alpha_a \pm \dfrac{\log_e 2}{T_a}} \tag{3.8}$$

Here α_a is the value of α at T_a; the plus sign applies to doubling the resistance and the minus sign to halving it.

3.2.1.4. VOLTAGE TURNOVER

The characteristic voltage turnover of an N.T.C. thermistor is shown in Fig. 1.2. The 'peak' or 'turnover' voltage can be obtained from Equation 3.2 by replacing R by V^2/P and then determining $\mathrm{d}V/\mathrm{d}P = 0$. This leads to

$$P_m^2 + P_m K(2T_a - B) + K^2 T_a^2 = 0 \tag{3.9}$$

where P_m is the power dissipation at turnover. Equation 3.9 may be solved to give

$$\frac{P_m}{K} = \frac{B}{2} - T_a \pm \frac{B}{2}\left(1 - \frac{4T_a}{B}\right)^{1/2} = T_m - T_a$$

where T_m is the corresponding thermistor temperature:

$$T_m = \frac{B}{2}\left[1-\left(1-\frac{4T_a}{B}\right)^{1/2}\right]$$

By using a binomial expansion for $(1-4T_a/B)^{1/2}$ we find that

$$P_m \approx \frac{1}{p_0}\left(1+\frac{2T_a}{B}\right) \approx \frac{1}{p_0} \tag{3.10}$$

and

$$T_m \approx T_a\left(1+\frac{T_a}{B}\right) \tag{3.11}$$

Since B lies between 1 500 and 6 000 K, T_m lies some 15–60 K above T_a, when T_a is room temperature.

3.2.5. Synthesis of the Static D.C. Characteristics[3]

If R_a, B and K are known for a thermistor mounted in a specified environment at ambient temperature T_a, then a family of approximate static current-voltage characteristics may be calculated using the relationships of Equations 3.1 and 3.2. This follows since $P = VI$ and $R = V/I$ are two simultaneous equations in V and I. A typical sequence of calculations would start with the known value of R_a at a specified T_a. T is then given a series of values. For each T, $T-T_a$ and TT_a are calculated, leading to $(T-T_a)/TT_a$, from which R is found from Equation 3.2. Then I^2 is found from P/R, P being known from Equation 3.1. Finally, V is determined from $V^2 = PR$. For most thermistors the variation of B with temperature is small enough to justify the assumption of B constant between the ambient temperature and the temperature T_m at the peak.

For points beyond the peak the assumption is questionable and it is more desirable to generate the IV curve from the experimentally-determined zero-power resistance-temperature data usually presented by the manufacturer. A typical form of presentation is shown in Fig. 2.1. The procedure in this case is as follows. Using log-log paper lines of constant resistance and constant power are plotted as shown for example in Fig. 1.2. A constant resistance line is selected and the corresponding $T-T_a$ found from the zero-power resistance-temperature curve for a given T_a. The power dissipation can then be found from Equation 3.1. A point on the characteristic is then given by the intersection of the corresponding constant power line with the constant resistance line initially chosen. The above procedure is repeated for a series of constant resistance lines and the complete characteristic generated for a given T_a. A family of characteristics is generated by choosing different values of T_a.

Synthesis of characteristics for constant T_a and variable K is similar. It is useful to realise, however, that for constant T_a the linear dependence of $T-T_a$ on P means that if one curve is available, for K_1, say, succeeding curves can be generated by sliding this curve bodily in the direction of the lines of constant R. This is possible since R is uniquely determined by T. The same procedure cannot be adopted in the generation of characteristics with K constant and T_a variable because of the non-linear dependence of resistance on temperature. Constant – T_a variable – K characteristics are shown in Fig. 1.3

A further set of static characteristics can be generated by extending Equation 3.1 to include the absorption of radiant power P_r by the thermistor. Then

$$P+P_r = K(T-T_a) \tag{3.12}$$

Such a family is shown in Fig. 1.4. To generate a point on a curve, T_a and P_r are chosen and a value of R is selected. R is a single-valued function of T and $T-T_a$ is therefore given by the resistance-temperature characteristic. Then, from Equation 3.12, $P = K(T-T_a)-P_r = V^2/R = I^2R$ can be found and hence the values of V and I.

3.2.6. Stability[4]

The working point for a thermistor fed from a Thévenin direct-voltage source E, with internal resistance S is determined by the intersection of the load line $V = E-IS$ with the static characteristic for T_a. Two possibilities can arise. In Fig. 3.1a load lines are shown which intersect the static characteristic at only one point. One working point is below turnover and the other above. These points are stable. It is, however, possible to have a load-line intersecting the static characteristic at three points as shown in Fig. 3.1b. The central point *B*, where the incremental resistance is negative, is an unstable one. This may readily be seen. Assume that, instantaneously, the operating

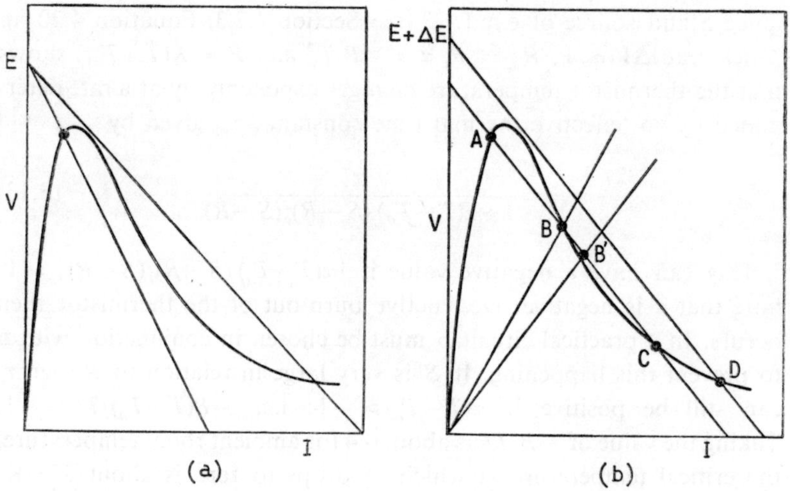

Fig. 3.1. Load line intersections with the static characteristics of N.T.C. thermistors; (a) single, (b) multiple

point is *B* and that for some reason there is a rise in *T*. This means that *R* falls and the operating point moves to *B'*. The power which must be dissipated in the thermistor is now such that its temperature rises further. Hence the dynamic working point rapidly moves to *C*. Similarly, a fall in *T*, while the thermistor is temporarily biased at *B* means a shift of operating point to *A*. Hence the working point on the characteristic curve will either by *A* or *C*, where the incremental, resistance is positive. The choice will be governed by the initial conditions. For example, if the available voltage at the source is *E* then the operating point will be *A* since the current will rise from zero to this value. If now the source voltage is increased steadily to $E+\Delta E$, as shown in Fig. 3.1b, the operating point will move to the peak point and then transfer very rapidly to the single point of intersection D in the negative resistance region. If the source voltage is then decreased, when the load line falls below the static characteristic the operating point shifts to the appropriate point of intersection in the positive resistance region.

Transient analysis[5–7] for an N.T.C. thermistor in series with a resistance S and source of e.m.f. E (see Section 4.2.3, Equation 4.20, in which read $\Delta V \ll V$, $R_\infty = R$, $\alpha = -B/T_w^2$ and $P = K(T-T_a)$) shows that the thermistor temperature changes exponentially at a rate determined by an 'effective thermal time constant', τ_e, given by

$$\tau_e = \frac{\tau}{1+\alpha(T-T_a)(S-R)/(S+R)} \tag{3.13}$$

This can have a negative value if $|\alpha(T-T_a)(S-R)/(S+R)| > 1$: note that α is negative. Destructive burn-out of the thermistor then occurs. In a practical circuit S must be chosen in conjunction with E to prevent this happening. If S is very large in relation to R then τ_e can still be positive if $\alpha(T-T_a) > -1$; i.e. $-B(T-T_a)/T^2 > -1$. Taking the value of $-B/T^2$ as about 0·4 for ambient room temperature, the critical temperature at which τ_e drops to zero is about 325 K. If the temperature is raised slightly above this by Joule heating, or by absorption of radiant power, burnout will occur.

3.2.7. Methods of Determining B and K[1,8]

The representation of R in Equation 3.2 readily leads to a determination of B and K from measured experimental values of R. Two approaches will be outlined. In the first R is measured for three values of applied power, which are also measured. Denoting these by the suffices 1, 2, 3, we obtain

$$\log_e R_1 - \log_e R_2 = \frac{B(P_2-P_1)/K}{(P_1/K+T_a)(P_2/K+T_a)} \tag{3.14}$$

$$\log_e R_2 - \log_e R_3 = \frac{B(P_3-P_2)/K}{(P_2/K+T_a)(P_3/K+T_a)} \tag{3.15}$$

Dividing and solving for K we find

$$K = \frac{1}{T_a} \frac{P_1(P_2-P_3)\log_e R_1+P_2(P_3-P_1)\log_e R_2+P_3(P_1-P_2)\log_e R_3}{(P_2-P_3)\log_e R_1+(P_3-P_1)\log_e R_2+(P_1-P_2)\log_e R_3} \tag{3.16}$$

K being known, B can be calculated from Equation 3.2.

The second approach is graphical. From Equations 2.8 and 2·9 we can write

$$\log_e R_a = \log_e R_c+B/T_a$$

$$\log_e R = \log_e R_c+B\bigg/\left(\frac{P}{K}+T_a\right)$$

so that

$$\left\{\log_e \frac{R_a}{R}\right\}^{-1} = \frac{T_a}{B}\left(1+\frac{T_aK}{P}\right) \tag{3.17}$$

Hence a plot of $\{\log_e (R/R_a)\}^{-1}$ against P^{-1} has a slope of T_aK and the intercept on the ordinate axis is T_a/B. A typical plot is shown in Fig. 3.2.

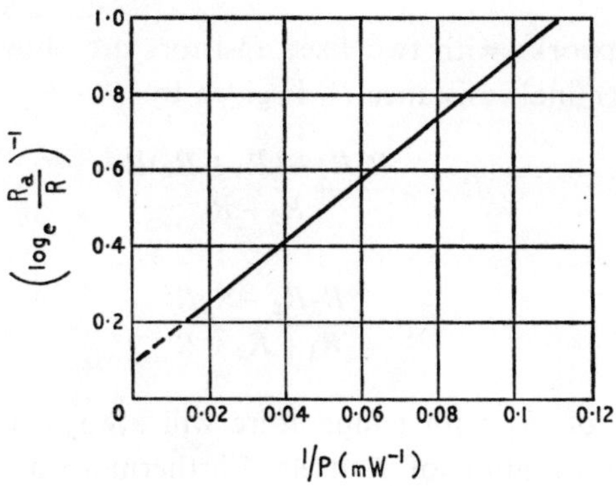

Fig. 3.2. Dependence of $(\log_e R_a/R)^{-1}$ *on* $1/P$ *for Stantel thermistor type A 1522/100*

3.2.8. Shaping of the Resistance-temperature Characteristic

In general the resistance-temperature characteristics of commercial thermistors do not have the form required for a specific application. Under certain conditions it is, however, possible to modify the characteristic so that it approximates a desired form. This is done using series-and shunt-connected resistors.[9, 10] If a simple design procedure is to be used the requirements are that the thermistor temperature is equal to the ambient temperature T_a and the passive resistors are also at T_a. It is also implicit that the thermistor obeys Ohm's Law. which is valid for N.T.C. thermistors at low power levels.

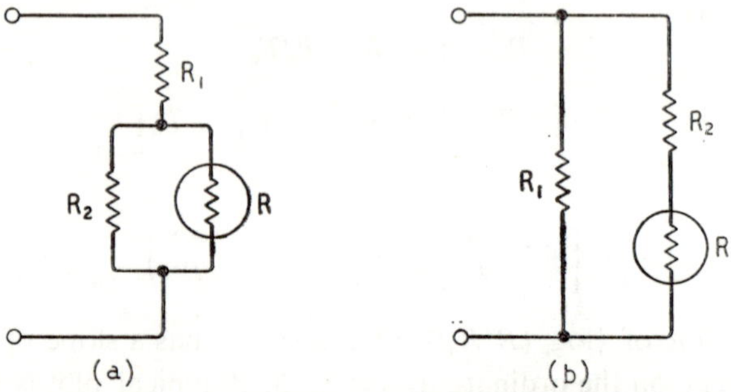

Fig. 3.3. Shaping networks with two fixed resistors

Shaping networks with two fixed resistors are shown in Fig. 3.3. For (a) the terminal resistance R_T is given by

$$R_T = \frac{R_1R_2+(R_1+R_2)R}{R_2+R} \tag{3.18}$$

and for (b)

$$R_T = \frac{R_1R_2+R_1R}{R_1+R_2+R} \tag{3.19}$$

The variation of R_T with temperature will always be less than the corresponding variation of R itself. Furthermore an increase in R will always cause an increase in R_T. These facts may be expressed mathematically as follows:

$$R_T = \frac{a+bR}{c+R} \tag{3.20}$$

with positive coefficients a, b and c. Geometrically, this corresponds to a rectangular hyperbola. More complicated networks involving more than two passive resistors can be used. Examples are shown in

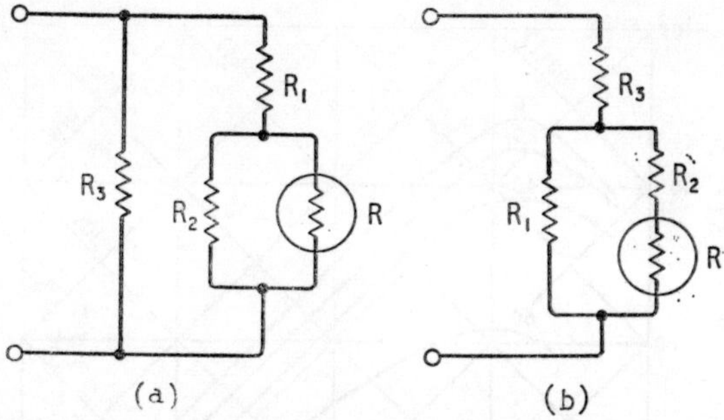

Fig. 3.4. More refined shaping networks with three fixed resistors

Fig. 3.4. These define a terminal resistance having the general form of Equation 3.18, but permit a better approximation to any desired curve. The synthesis procedure for a required R_T/R relationship is considered by Keonjian and Schaffner[9].

The most common requirement is a near-linear dependence of current on temperature in temperature measurement applications. This is discussed in detail in Section 8.2.1.

3.2.9. The Critical Temperature Thermistor

This is the type of oxide thermistor based on Vd_2O_5, which shows an extremely large negative temperature coefficient of resistance in the region of 68 °C bead temperature. There is no simple analytical form for the $V-I$ characteristics[11], but a typical set of characteristics with T_a as parameter is shown in Fig. 3.5.

3.3. P.T.C. THERMISTORS

3.3.1. Barium Titanate Based Types

Saburi and Wakino[12] have used Equations 3.1 and 2.17 to specify $V-I$ characteristics of 'Posistors' having a positive temperature coefficient of resistance over the whole range of operation in the manner

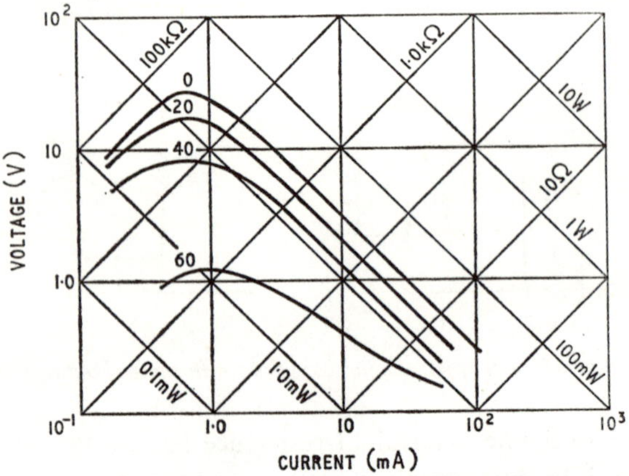

Fig. 3.5. Typical I/V characteristics of the critical temperature resistor. The parameter is ambient temperature in °C. (After Futaki)

of Section 3.2.5. They deduce the following relationships at the current maximum.

$$\left.\begin{aligned} I_m &= \frac{K}{AR_c}\exp(AT_a-1)^{1/2}; \quad V_m = \frac{KR_c}{A}\exp(AT_a+1)^{1/2} \\ P_m &= K/A; \quad T_m = 1/A+T_a \\ R_m &= R_c\exp(AT_a+1) \end{aligned}\right\} \qquad (3.21)$$

Practical characteristics of their thermistors are shown in Fig. 3.6 and that for a normal N.T.C. characteristic is included for comparison. They note, however, the basis of this approach—that R depends only on T—is an approximation. The thermistor resistance is also voltage-dependent, particularly in the para-electric phase, namely in the temperature region above T_p.

Andrich[13] has shown that the $V-I$ characteristic can be well represented by that of the parallel combination of a pure thermistor (temperature-dependent resistor) and a varistor (voltage-dependent resistor). This is shown in Fig. 3.7. Here curve (a) is that of the practical thermistor (showing a second turnover at high V); it can be synthesised by combining curves (b) and (c); curve (b) is calculated for

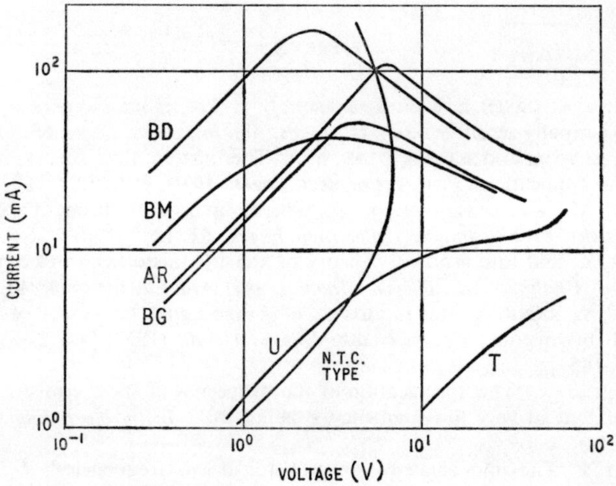

Fig. 3.6. Static characteristics of 'posistors' (after Saburi and Wakino)

the 'pure thermistor' $(R-T)$ from Equation 3.1 and $P = I^2R = V^2/R$; curve (b) is the difference between (a) and (b) and is attributed to a voltage-dependent resistor.

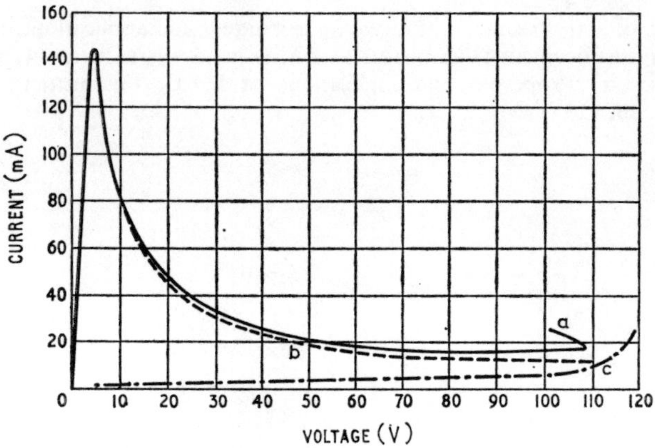

Fig. 3.7. Static characteristic of P.T.C. thermistor 2322 660 90003 (solid curve) This is synthesised by addition of theoretical (dashed curve) based on contact linearity and end-contact (dot-dashed) curve. (after Andrich) ——— P.T.C. experimental, - - - - - - P.T.C. theoretical, ·—·—·— V.D.R.

REFERENCES

1. BECKER, J. A., GREEN, C. B., and PEARSON, G. L. 'Properties and uses of thermistors—thermally-sensitive resistors', *Trans. Am. Inst. elect. Engrs.* **65,** 711 (1946)
2. SCARR, R. W. A. and SETTERINGTON, R. A. 'Thermistors, their theory, manufacture and application', *Proc. Instn. elect. Engrs.*, **107B,** 395, 408 (1960)
3. SAPOFF, M., and OPPENHEIM, R. M. 'Theory and application of self-heated thermistors' *Proc. Instn. elect. electronic Engrs.*, **51,** 1292 (1963)
4. EKELÖF, S., and KIHLBERG, G. 'Theory of the thermistor as an electric circuit element', *Chalmers tek. Högsk. Handl.* (1945) *Avd Elektrotevnik* 40, Nr. 142
5. EKELÖF, S., BJÖRK, N. and DAVIDSON, R. 'Large-signal behaviour of directly-heated thermistors', *Chalmers tek. Högsk. Handl.* (1957) *Avd Elektroteknik* 56, Nr. 185
6. CANDY, C. J. N. 'The specification of the properties of the thermistor as a circuit element in very low-frequency systems' *Proc. Instn. elect. Engrs.*, **103B,** 398 (1954)
7. HYDE, F. J. 'The impedance of a thermistor at low frequencies', *J. Electron.*, **1,** 303 (1955)
8. BJÖRK, N. and DAVIDSON, R. 'Small-signal behaviour of directly-heated thermistors', CHALMERS *tek. Högsk. Handl.*, (1955) *Avd. Elektroteknik* 48 Nr 169
9. KEONJIAN, E. and SCHAFFNER, J. S. 'Shaping of the characteristics of temperature-sensitive elements', *I.R.E. Trans. compon. Parts*, **4,** 396 (1954)
10. SOBLE, A. B. 'Bounds for thermistor compensation of resistance and conductance', *I.R.E. Trans. compon. Parts.*, **7,** 96 (1957)
11. FUTAKI, H. 'A new type semiconductors critical temperature resistor', *J. Appl. Phys. Japan*, **4,** 28 (1956)
12. SABURI, O. and WAKINO, K. 'Processing techniques and applications of positive temperature coefficient thermistors', *I.R.E. Trans. compon. Parts*, **14,** 53 (1963)
13. ANDRICH, E. 'Properties and applications of P.T.C. thermistors', *Electron. Applic.* **26,** 123 (1966)

CHAPTER 4

A.C. and Transient Characteristics

4.1. INTRODUCTION

The thermal inertia, specified by the thermal time constant, τ, gives rise to interesting and useful relationships between terminal current and voltage. It is convenient to distinguish between low-and high-level conditions. In the former, the applied excitation, (for example a sine wave or step function of current or voltage), is small in relation to the biasing parameters, but in the latter is not. For low-level conditions small-signal electrical circuits can be derived.

Both small-signal and large-signal conditions are treated in this chapter and comparisons are made between theory and experiment.

4.2. LOW-LEVEL ANALYSIS

4.2.1. General Small-signal Analysis

Small-signal equivalent circuits for biased thermistors were derived by Ekelöf and Kihlberg[1] in 1954 and independently by Burgess[2] in 1955. Burgess' treatment will be followed here.

Let V and I be the instantaneous voltage and current applied to the thermistor, whose instantaneous temperature is T. The initial assumption made is that T is single valued and uniform throughout the device, so that internal temperature gradients, such as will arise to some degree in practice, are ignored.

If the rate of loss of heat is determined solely by the instantaneous excess temperature, $\theta = T - T_a$, above the ambient temperature T_a, then the balance-of-power equation is

$$C\frac{d\theta}{dt} + f(\theta) = P = IV \tag{4.1a}$$

which reduces to

$$C\frac{d\theta}{dt} + K\theta = IV \tag{4.1b}$$

for Newtonian cooling, which will be assumed in this section. In these equations C is the heat capacity of the thermistor and will be assumed independent of T, and K is the dissipation constant or thermal conductance.

I, V and θ will be assumed to have steady components, with relatively small a.c. components at angular frequency ω superposed, as follows:

$$I = I_0 + I_1 \exp(j\omega t) \tag{4.2a}$$

$$V = V_0 + V_1 \exp(j\omega t) \tag{4.2b}$$

$$T - T_a = \theta = \theta_0 + \theta_1 \exp(j\omega t) \tag{4.2c}$$

The mean current and voltage, I_0 and V_0, respectively, define a point on the steady-state characteristic for the ambient temperature T_a: this point is also on the isothermal curve for temperature $T_a + \theta_0$. The quantities I_1, V_1 and θ_1 are in general complex, carrying both amplitude and phase information. The corresponding expression for instantaneous power is, approximately,

$$P = P_0 + P_1 \exp(j\omega t) = IV = I_0V_0 + (I_0V_1 + I_1V_0) \exp(j\omega t) \tag{4.3}$$

if $|I_1|$, $|V_1| \ll I_0$, V_0, respectively. Substitution from Equation 4.2c and 4.3 into Equation 4.1 gives

$$K\theta_0 = I_0V_0 \tag{4.4a}$$

$$\theta_1(K + j\omega C) = I_0V_1 + I_1V_0 = P_1 = \theta_1 y \tag{4.4b}$$

Here $y = P_1/\theta_1$ is the *thermal admittance*. This may be expressed as

$$y = \frac{C}{\tau} + j\omega C \tag{4.5}$$

where $\tau = C/K$ is the thermal time constant of the thermistor; i.e. the time taken for the excess temperature to fall to $1/e$ of its initial value if the d.c. and a.c. sources are removed. Now the alternating component of current I_1 is determined by both V_1 and θ_1 through the isothermal a.c. conductance g and the current-temperature coefficient, h:

$$I_1 = \left(\frac{\partial I}{\partial V}\right)_T V_1 + \left(\frac{\partial I}{\partial T}\right)_V \theta_1 = gV_1 + h\theta_1 \tag{4.6}$$

The sign of h distinguishes N.T.C. and P.T.C. thermistors: $h > 0$ signifies N.T.C. and $h < 0$, P.T.C. The *a.c. admittance* is

$$Y(\omega) = \frac{I_1}{V_1} = \frac{gy + hI_0}{y - hV_0} \tag{4.7}$$

At zero frequency this becomes a pure conductance

$$Y_0 = G_0 = \frac{gK + hI_0}{K - hV_0} \tag{4.8}$$

G_0 is simply the slope of the steady-state current-voltage characteristic. Another way of looking at this is to consider small and infinitely-slow increments ΔI and ΔV along the steady state characteristic, accompanied by a temperature change $\Delta\theta$, so that

$$K\,\Delta\theta = I_0\,\Delta V + V_0\,\Delta I \tag{4.9a}$$

$$\Delta I = g\,\Delta V + h\,\Delta\theta \tag{4.9b}$$

These Equations may be combined to give $G_0 = \Delta I/\Delta V$, in the form of Equation 4.7.

At infinite frequency the a.c. admittance is

$$Y_\infty \equiv G_\infty = g \tag{4.10}$$

This is again purely real because the instantaneous temperature of the device cannot follow the infinitely rapid variation of electrical power.

The general form of admittance, $Y(\omega) = G(\omega)+jB(\omega)$, may be expressed in terms of $G(\omega)$ and $B(\omega)$ as follows:

$$G(\omega) = \frac{g\omega^2C^2+(gC/\tau+hI_0)(C/\tau-hV_0)}{\omega^2C^2+(C/\tau-hV_0)^2}$$

$$= g+\frac{h(I_0+gV_0)(C/\tau-hV_0)}{\omega^2C^2+(C/\tau-hV_0)^2} \quad (4.11a)$$

$$B(\omega) = -\frac{\omega Ch(gV_0+I_0)}{\omega^2C^2+(C/\tau-hV_0)^2} \quad (4.11b)$$

The corresponding formulae for the impedance representation,

$$Z(\omega) = 1/Y(\omega) = R(\omega)+jX(\omega), \quad \text{are}$$

$$R(\omega) = \frac{g\omega^2C^2+(C/\tau-hV_0)(gC/\tau+hI_0)}{g^2\omega^2C^2+(gC/\tau+hI_0)^2} \quad (4.12a)$$

$$X(\omega) = \frac{\omega Ch(gV_0+I_0)}{g^2\omega^2C^2+(gC/\tau+hI_0)^2} \quad (4.12b)$$

Using the identity $g \equiv I_0/V_0$, *which is valid for linear isothermals*, which arise for N.T.C. thermistors, it follows from Equation 4.8 that

$$hI_0 = \frac{gK(G_0-g)}{(G_0+g)}$$

$$hV_0 = \frac{K(G_0-g)}{(G_0+g)}$$

Making use of the identities $G_0 \equiv 1/R_0$, $g \equiv 1/R_\infty$ and $\tau = C/K$, together with the above expressions for hI_0 and hV_0, we obtain alternative forms for Equations 4.11 and 4.12, as follows:

$$G(\omega) = \frac{\dfrac{\omega^2\tau^2}{R_\infty}+\dfrac{4R_0}{(R_0+R_\infty)^2}}{\omega^2\tau^2+\dfrac{4R_0^2}{(R_0+R_\infty)^2}} \quad (4.11c)$$

$$B(\omega) = -\frac{\dfrac{2\omega\tau(R_\infty-R_0)}{R_\infty(R_\infty+R_0)}}{\omega^2\tau^2+\dfrac{4R_0^2}{(R_0+R_\infty)^2}} \quad (4.11d)$$

$$R(\omega) = \frac{R_0 + \dfrac{\omega^2\tau^2(R_\infty + R_0)^2}{4R_\infty}}{1 + \dfrac{\omega^2\tau^2(R_\infty + R_0)^2}{4R_\infty^2}} \tag{4.12c}$$

$$X(\omega) = \frac{\omega\tau(R_\infty^2 - R_0^2)/2R_\infty}{1 + \dfrac{\omega^2\tau^2(R_\infty + R_0)^2}{4R_\infty^2}} \tag{4.12d}$$

Both the impedance and admittance loci, when plotted in the complex plane, are semicircles with diameters on the real axis, if g, C, h and τ are independent of frequency (see, however, Section 4.2.4). This is clear if, for example, the expression for $Z(\omega)$ is rewritten in the form

$$Z(\omega) = R_\infty - \frac{(R_\infty - R_0)}{1 + j\omega\tau\left(\dfrac{R_\infty + R_0}{2R_\infty}\right)} \tag{4.13}$$

If G_0 is negative, then at a critical frequency defined by

$$\omega_0^2 = \left(\frac{hV_0}{C} - \frac{1}{\tau}\right)\left(\frac{hI_0}{gC} + \frac{1}{\tau}\right) = \frac{4gG_0}{\tau^2(G_0+g)^2} \tag{4.14}$$

the terminal conductance will be zero and I_1 and V_1 will be in phase-quadrature. This situation is depicted by the admittance locus on the left of Fig. 4.1 for an N.T.C. thermistor. From the geometry it is apparent that the susceptance at ω_0 is given by

$$B(\omega_0) = (-G_0 g)^{1/2}$$

The condition that G_0 is negative is $h > C/\tau V_0$ for an N.T.C. thermistor and $h < -gC/\tau I_0$ for a P.T.C. thermistor. The right-hand locus in Fig. 4.1 is for an N.T.C. thermistor biased below turnover, so that G_0 is positive,

It is instructive to consider the load line approach for small-signal a.c. operation. At very low frequencies, such that $\omega \ll 1/\tau$, the temperature is cophasal with IV and the locus of the operating point is a small arc of the static characteristic (straight line for the small-signal case being considered, corresponding to the slope conductance, G_0).

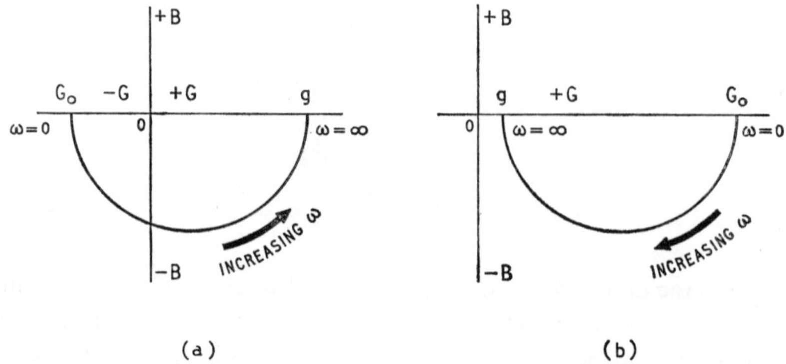

Fig. 4.1. Small-signal admittance loci for an N.T.C. thermistor biased (a) beyond turnover, (b) below turnover

At high frequencies, such that $\omega \gg 1/\tau$, the power variation is so rapid that the temperature cannot change and isothermal conditions prevail. The locus is then a section of the isothermal characteristic. In the intermediate frequency range the instantaneous temperature lags the instantaneous power dissipation with a phase-angle corresponding to that of the thermal admittance y, so giving rise to an approximately elliptical locus as shown in Fig. 4.2a. The major axis of the ellipse may have either a positive or negative slope according to the amplitude

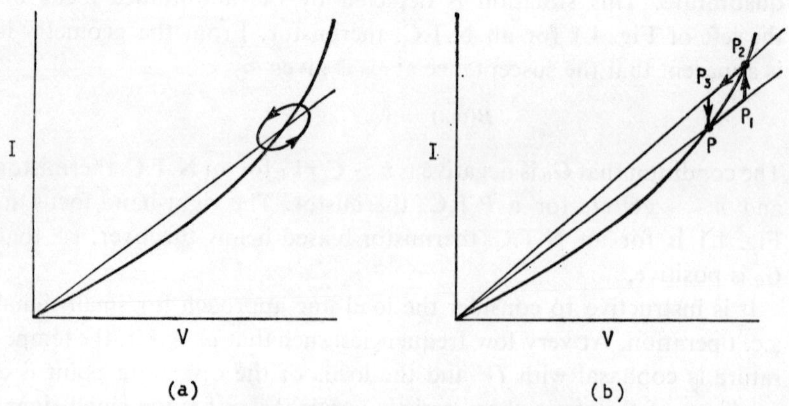

Fig. 4.2. Locus of the operating point for (a) sinusoidal and (b) square-wave excitation of an N.T.C. thermistor

and frequency of the excitation. It will be noted that the direction in which the ellipse is traced out is anticlockwise, for bias beyond turn-over. That this is so may be seen by reference to Fig. 4.2b. If a small instantaneous departure from the established operating point is produced by a voltage change then initially the thermistor temperature is as before and the operating point follows the ohmic line from the origin from P to P_1. As the thermistor warms up so its resistance falls and the operating point moves to P_2. If the small voltage excitation is removed then the operating point moves via P_3 back to P. The lead of voltage change over current change is characteristic of the inductive behaviour already established. For P.T.C. thermistors the converse is true and the locus is traced in a clockwise direction if I and V are plotted as in the same way. At the critical frequency such that the admittance is a pure susceptance the current lags the voltage by $\pi/2$ radians and the axes of the ellipse are parallel to the I and V axes, respectively.

4.2.2. Small-signal Equivalent Circuits[1–3]

Equivalent circuits corresponding to the analytical expressions of Equations 4.11 and 4.12 are shown in Fig. 4.3. In the upper part of the figure two forms are shown for N.T.C. thermistors ($h > 0$, $G_0 > g$) and below, two corresponding equivalent circuits for P.T.C. thermistors ($h < 0$, $G_0 < g$). It is convenient to specify the elements in terms of R_0, R_∞ and τ, since these may readily be determined experimentally. R_0 and R_∞ are very simply obtained from d.c. measurements. τ can be calculated from the measured *effective circuit time constant* introduced in the next section.

In some applications in which the operating point is defined by an external source of e.m.f. and a series resistor, changes in it can be brought about by changes in ambient temperature T_a and/or dissipation constant K.[4] When such changes are relatively small an appropriate small-signal equivalent circuit can be readily devised *if* the changes in K or T_a are very slow in relation to $1/\tau$.

The form of the equivalent circuit is similar to those for the vacuum

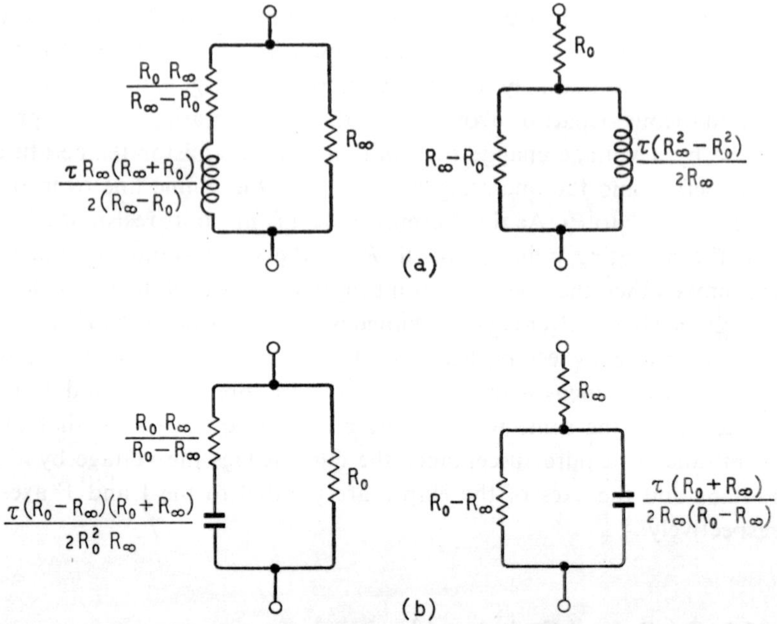

Fig. 4.3. Small-signal equivalent circuits of thermistors; (a) N.T.C., (b) P.T.C.

triode. It is shown in Fig. 4.4. The resistance, R_0, is simply the slope of the static characteristic at the operating point, formally defined here again as

$$R_0 = \left.\frac{\delta V}{\delta I}\right|_{K,\,T\ \text{const}}.$$

The 'amplification' factor is comprised of two parts, in general, since excitations of both dissipation factor and ambient temperature are possible. For N.T.C. thermistors (current-controlled devices), these are defined respectively as

$$\mu_K = \left.\frac{\delta V}{\delta K}\right|_{I,\,T_a\ \text{const}}; \qquad \mu_{T_a} = \left.\frac{\delta V}{\delta T_a}\right|_{I,\,K\ \text{const}}$$

It is obvious from the characteristics of Figs. 1.3 and 1.2a respectively, that μ_K is positive and μ_{T_a} is negative. μ_K can be derived by differen-

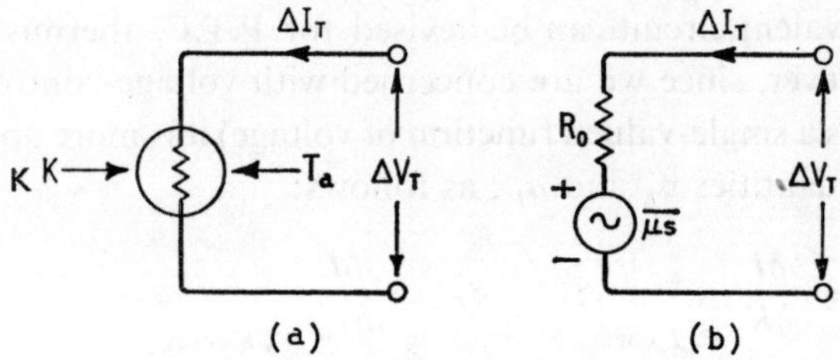

(c) $\Delta V_T = R_0 \Delta I_T + \overline{\mu s}$ (s = STIMULUS)

(d) $\overline{\mu s} = \mu_K \Delta K + \mu_{Ta} \Delta T_a$

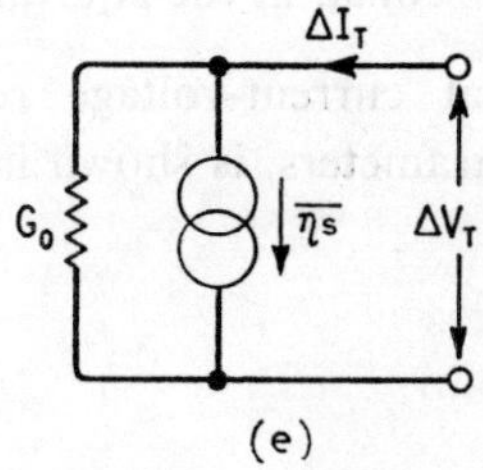

(f) $\Delta I_T = G_0 \Delta V_T + \overline{\eta s}$

(g) $\overline{\eta s} = \eta_K \Delta K + \eta_{Ta} \Delta T_a$

4.4 Small-signal representation of thermistors (a–d, N.T.C., e–g, P.T.C.): (a) circuit; (b) equivalent circuit; (c) small-signal equation; (d) definition of $\overline{\mu s}$; (e) equivalent circuit; (f) small-signal equation; (g) definition of $\overline{\eta s}$

tiating the two expressions for V, $V = IR$ and $V = K(T-T_a)/I$ with R defined as in Equation 2.9;

$$\frac{\delta V}{\delta K} = -\frac{BV}{T^2}\frac{\mathrm{d}T}{\mathrm{d}K} = \frac{T-T_a}{I} + \frac{K}{I}\frac{\mathrm{d}T}{\mathrm{d}K}$$

Elimination of $\mathrm{d}T/\mathrm{d}K$ gives

$$\mu_K = \frac{V}{1-K/\alpha P}\cdot\frac{1}{K} \tag{4.15a}$$

where α as defined in Equation 2.11 is negative. Similarly it can be shown that

$$\mu_{T_a} = -\frac{V}{1-K/\alpha P}\cdot\frac{1}{T-T_a} \tag{4.15b}$$

The overall relation between the small-signal quantities is shown in Fig. 4.4.

A similar equivalent circuit can be devised for P.T.C. thermistors. In this case, however, since we are concerned with voltage-controlled devices (current is a single-valued function of voltage) it is more appropriate to define quantities η_K and η_{T_a}, as follows:

$$\eta_K = \left.\frac{\delta I}{\delta K}\right|_{V,\,T\ \text{const}} ; \qquad \eta_{T_a} = \left.\frac{\delta I}{\delta T_a}\right|_{V,\,K\ \text{const}}$$

Here η_K is positive and η_{T_a} negative. It is also more appropriate to use the slope conductance, G_0, of the static characteristic at the operating point, defined as $(\delta I/\delta V)\big|_{K,\,T_a}$ const, in the equivalent circuit. This, and the terminal small-signal current-voltage relationship, expressed in terms of small-signal parameters, is shown in Fig. 4.4b.

4.2.3. Low-level Transient Analysis

We consider a simple series circuit comprising a thermistor, a resistor of resistance S and battery of e.m.f. V with an additional series e.m.f. $\Delta V \ll V$, which can be switched in series with V. In the initial steady-state condition determined by V, we can adapt Equation 4.1b to obtain

$$K(T_w - T_a) = V^2 \frac{R_\infty}{(R_\infty + S)^2} \tag{4.16}$$

where T_w is the thermistor temperature and R_∞ is the corresponding d.c. (isothermal) resistance. On applying the step of e.m.f., ΔV, the equation defining the transient conditions is

$$C\frac{\mathrm{d}T}{\mathrm{d}t} + K(T - T_w) = (V + \Delta V)^2 \frac{R'_\infty}{(R_\infty + S)^2} - V^2 \frac{R_\infty}{(R_\infty + S)^2} \tag{4.17}$$

For small temperature differences, $T - T_w$, we may suppose that R'_∞, the instantaneous value of R_∞, has the form

$$R'_\infty = R_\infty\{(1 + \alpha(T - T_w))\} \tag{4.18}$$

Here $\alpha = -B/T_w^2$ is the temperature coefficient of resistance at the working point (see Equation 3.6). Substitution for R'_∞ in Equation 4.17 and neglect of second-order terms in $T-T_w$ leads to

$$C\frac{\mathrm{d}T}{\mathrm{d}t}+(T-T_w)\left\{K-(V+\Delta V)^2\frac{R_\infty}{(R_\infty+S)^2}\frac{B}{T_w^2}\frac{(R_\infty-S)}{(R_\infty+S)}\right\}$$

$$=(V^2+2V\,\Delta V)\frac{R_\infty}{(R_\infty+S)^2} \tag{4.19}$$

By inspection it can be seen that the *effective thermal time-constant* for the circuit is

$$\tau_e=\frac{\tau}{\left\{1-(V+\Delta V)^2\dfrac{R_\infty}{(R_\infty+S)^2}\dfrac{B}{KT_w^2}\dfrac{(R_\infty-S)}{(R_\infty+S)}\right\}} \tag{4.20}$$

We have assumed that for small temperature differences $R'_\infty-R_\infty$ is linearly related to $T-T_w$, so that the time constant governing the rate of change of R'_∞ is likewise τ_e. Furthermore, since the changes in R'_∞ are small, the rate of change of current is also governed by τ_e, which may therefore be formally identified with the *effective time constant for the circuit*, τ_c.

τ_c specifies the rate of change of resistance with time and is the parameter of importance for circuit applications. This will now be derived for an N.T.C. thermistor. With the nomenclature of this section we have the two basic equations relating I and V;

$$P=IV=I^2R_\infty=V^2/R_\infty=K(T_w-T_a) \tag{1.1a}$$

$$R_\infty=V/I=R_c\exp(B/T_w) \tag{2.8b}$$

From the total differential of Equation 1.1a we obtain

$$\frac{\mathrm{d}V}{\mathrm{d}I}=\frac{K}{I_w}\frac{\mathrm{d}T}{\mathrm{d}I}-\frac{V_w}{I_w}$$

and from the corresponding differential of Equation 2.8b,

$$\frac{\mathrm{d}V}{\mathrm{d}I}-\frac{V\,\mathrm{d}I}{I_w^2}=\frac{-B}{T_w^2}\frac{V_w}{I_w}\,\mathrm{d}T$$

Combining these leads to

$$\frac{dV}{dI} = \frac{KT_w^2}{V_w B I_w}\left(\frac{V_w}{I_w} - \frac{1}{I_w}\frac{dV}{dI}\right) - \frac{V_w}{I_w}$$

which may be put into the form

$$\frac{KT_w^2}{BI_w^2 R_\infty} = \frac{R_\infty + R_0}{R_\infty - R_0}$$

and because

$$I_w = \frac{V_1}{R_\infty + S} \simeq \frac{V_1 + \Delta V}{R_\infty + S}$$

this becomes

$$\frac{KT_w^2(R_\infty + S)^2}{B(V_1 + \Delta V)^2 R_\infty} = \frac{R_\infty + R_0}{R_\infty - R_0} \qquad (4.21)$$

Substitution in Equation 4.19 yields

$$\tau_c = \tau \frac{(R_\infty + R_0)}{2R_\infty}\frac{(R_\infty + S)}{(R_0 + S)} \qquad (4.22)$$

For $S = \infty$ Equation 4.22 reduces to

$$\tau_c = \tau_e \equiv \tau \frac{(R_\infty + R_0)}{2R_\infty} \qquad (4.23)$$

which is a property of the thermistor itself.

If R_0 and R_∞ are known and τ_e is measured for a range of values of S, τ can be calculated.

Equation 4.22 may be derived directly after using either of the equivalent circuits in the upper part of Fig. 4.3 to represent the thermistor in the circuit which has been discussed.

.42.4. Non-Newtonian Cooling

Newtonian cooling requires that the rate of loss of heat is a function only of the excess temperature $T - T_a$ and does not depend on the derivatives $\delta(T-T_a)/\delta t$, $\delta^2(T-T_a)/\delta t^2$, etc. It also implies that the thermal conductance K is independent of frequency. Following Burgess, the effect of departure from Newtonian cooling will be illustrated by considering the case of heat loss by diffusion in one dimension.

Let $x = 0$ represent the centre of the thermistor and $x = l$ the distance from the centre at which the device temperature (lead wire) is at the ambient temperature T_a. The flow of heat is governed by Fick's Second Law:

$$D\frac{\partial^2\theta}{\partial x^2} = \frac{\partial\theta}{\partial t} \tag{4.24}$$

where the thermal diffusivity $D = K/\varrho s$; ϱ is the density and s the specific heat of the conducting path.

The expression for the excess temperature at x has the form

$$\begin{aligned}\theta(x) &= \theta_0(x)+\theta_1(x)\exp \mathrm{j}\omega t \\ &= \theta_0\left(\frac{l-x}{l}\right)+\{A\exp(ax)+B\exp(-ax)\}\exp \mathrm{j}\omega t\end{aligned} \tag{4.25}$$

From the boundary conditions at $x = 0$ and $x = l$ it follows that the instantaneous power at $x = 0$ is given by

$$P = P_0+P_1\exp \mathrm{j}\omega t = \frac{KX\theta_0}{l}+\theta_1(\mathrm{j}\omega C+aKX\coth al)\exp \mathrm{j}\omega t \tag{4.26}$$

where X is the cross-sectional area of the conducting path.

It can be seen, by comparing this with Equation 4.4b, that $aXK \coth al$ has replaced K in the expression for the thermal admittance P_1/θ_1. Only if $|al|$ is small compared with unity is this term independent of frequency and then $l(\omega/D)^{1/2} \ll 1$, so that

$$aXK\coth al = \frac{XK}{l} = \frac{P_0}{\theta_0} \tag{4.27}$$

which is simply the d.c. thermal conductance of the cooling path. On the other hand if l or ω are sufficiently large $l(\omega/D)^{1/2} \gg 1$, so that

$$aXK\coth al = (1+\mathrm{j})(\omega/2D)^{1/2}XK \tag{4.28}$$

which is complex and increases as $\omega^{1/2}$. This means that the thermal time constant τ of Equation 4.5 can no longer be regarded as a real constant in Equations 4.11 and 4.12, although the general relations in Equations 4.7, 4.8 and 4.10 are still valid.

Taking the thermal admittance y as $(j\omega C+aXK \coth al)$ leads to the following expression for electrical admittance

$$Y(\omega) = \frac{g(j\omega C+aXK \coth al)+hI_0}{(j\omega C+aXK \coth al)-hV_0} \tag{4.29}$$

It follows from Equation 4.7 that, whatever the nature of the thermal admittance, y;

$$\frac{G_0-Y(\omega)}{Y(\omega)-g} = \frac{y-y_0}{y_0-hV_0} \tag{4.30}$$

where y_0 is the thermal conductance at zero frequency. Hence, if $y-y_0$ has a constant phase angle, ϕ, at all frequencies the locus of $Y(\omega)$ will be circular. Also the chord (g, G) will subtend an angle 2ϕ at the centre of the circle. This follows from the geometry of the angles subtended by the chord of a circle at its centre. For one-dimensional diffusion

$$y-y_0 = j\omega C+aXK(\coth al-1/al) \tag{4.31}$$

clearly does not have a constant phase angle. Hence the loci of $Y(\omega)$ and $Z(\omega)$ will not be circular.

The application of Equation 4.30 to the simple case of Newtonian cooling, where $Y(\omega) = j\omega C+K\theta$ and $y-y_0$ has a phase-angle $\pi/2$ at all frequencies, shows that the locus is a circle with g, G_0 as the ends of the diameter.

4.2.5. Practical Considerations

Impedance loci have been determined experimentally by Candy[6], Björk and Davidson[7], and Hyde[3]. Candy used an electrically driven motor-generator set to obtain the low frequencies required and amplitude and phase balance were separate. Björk and Davidson determined terminal impedance from an elliptical current-voltage display obtained conventionally, using an electronic low-frequency, sinusoidal signal source. Hyde also used an electronic signal source and formed a Maxwell bridge in which the thermistor occupied one arm: in view of the low frequencies which were involved stray reactances were not troublesome. Typical results are shown in Fig. 4.5a for a bead N.T.C. thermistor type A5412, at an ambient temperature of 303 K. The three loci in Fig. 4.5 are for bias currents of 0·5 mA, 1·0 mA and 2·0 mA, the last two being beyond turnover, so that R_0

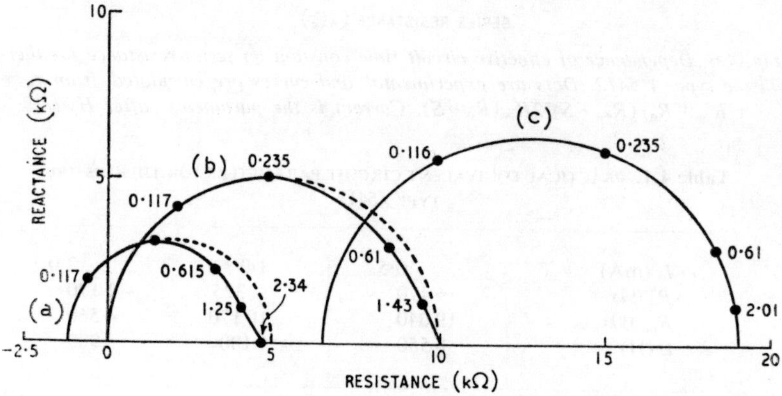

Fig. 4.5. Experimental small-signal impedance loci for an N.T.C. thermistor type A 5412; bias currents are (a) 2·0 mA, *(b) 1·0* mA, *(c) 0·5* mA. *Frequency is in* Hz. *(after Hyde)*

is negative. It will be noted that there is a slight departure from the semicircle at higher frequencies, as is predicted by the analysis of the previous section.

The thermal time constants of thermistors have been determined from transient measurements involving out-of-balance signals in Wheatstone bridge circuits when the supply voltage was suddenly changed. As shown previously the circuit time constant is jointly determined by the thermistor and the resistance S in series with it (see Equation 4.22), so that the determination of τ requires a knowledge of S, R_0 and R_∞. In Fig. 4.6 curves of τ_e versus S are shown for the same thermistor for which the results of Fig. 4.5 apply. Corresponding theoretical points utilising a derived value for τ of 1.4 s are also shown: these agree well with experimental values. Equivalent

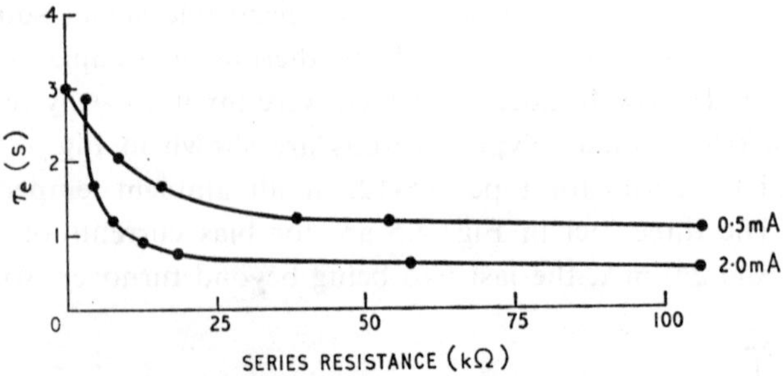

Fig. 4.6. Dependence of effective circuit time constant on series resistance for thermistor type A 5412. Dots are experimental and curves are calculated from $\tau_e = \tau(R_\infty + R_0)(R_\infty + S)/2R_\infty(R_0 + S)$. *Current is the parameter (after Hyde)*

Table 4.1. PRACTICAL EQUIVALENT CIRCUIT PARAMETERS FOR THERMISTOR TYPE A5412

I_0 (mA)	0·5	1·0	2·0
R_0 (Ω)	6 470	− 325	− 1 170
R_∞ (Ω)	19 040	10 170	4 560
L (H)	8 550	7 100	2 980

circuit data for this thermistor are given in Table 4.1 for the bias currents used. Derived data of this kind have also been published by Ekelöf and Kihlberg.

It is evident that the choice of a suitable operating point can result in the above thermistor, for example, exhibiting an inductance of several thousand henries—much larger than can be obtained with wound coils. From the practical point of view the equivalent Q-factor of the inductor, which the thermistor represents electrically, in conjunction with an effective shunt resistance S due to the bias circuit, is important.[3] The Q-factor in question is the ratio of the magnitudes of the equivalent series reactance and resistance components associated with the parallel combination of thermistor and S. For a given working point, defining R_0 and R_∞, this is

$$Q = \frac{\omega LS(R_\infty - R_0)^2}{\omega^2 L^2 R_\infty (R_\infty + S) + R_0(R_0 + S)(R_\infty - R_0)^2} \tag{4.32}$$

This has a maximum value $\hat{Q}$ with respect to frequency when the two terms in the denominator are equal. The corresponding frequency is

$$\hat{\omega} = \frac{R_\infty - R_0}{L}\left\{\left(\frac{S+R_0}{S+R_\infty}\right)\frac{R_0}{R_\infty}\right\}^{1/2} \quad (4.33)$$

and

$$\hat{Q} = \frac{S(R_\infty - R_0)}{2}\left\{\frac{1}{R_0 R_\infty (S+R_0)(S+R_\infty)}\right\}^{1/2} \quad (4.34)$$

Q is slowly-dependent on frequency in the vicinity of $\hat{\omega}$, relative to the linear dependence or inverse dependence which arise according as ω^2 is either much smaller or much greater than $R_0(S+R_0)(R_\infty - R_0)^2/L^2R_\infty(R_\infty+S)$. As turnover is approached from low currents, $Q \to \infty$ for $R_0 = 0$ but at the same time $\hat{\omega} \to O$. $\hat{Q}$ will also reach an upper limiting value with respect to S as $S \to \infty$. This is

$$\hat{Q}_s = (R_\infty - R_0)/2(R_0 R_\infty)^{1/2} \quad (4.35)$$

For the type A5412 transistor with $I_0 = 0{\cdot}5$ mA we find that

$$\hat{Q}_s = 0{\cdot}56$$
$$\omega = 3{\cdot}85 \text{ rad/s } (0{\cdot}61 \text{ Hz})$$

4.3. HIGH-LEVEL ANALYSIS FOR NEWTONIAN COOLING[5]

Provided that the instantaneous resistance of the thermistor can be specified to a good approximation in a relatively simple closed form, then it is possible to obtain approximate predictions of circuit behaviour under large-signal conditions.

4.3.1. A.C. Signals

We will consider a simple series circuit at ambient temperature T_a, comprising a sinusoidal source of e.m.f. $e(t) = E \sin \omega t$, a resistance S and an N.T.C. thermistor, whose instantaneous resistance is specified as

$$R(T) = R_c \exp(B/T) \tag{4.36a}$$

or

$$R(\theta) = R_c \exp\{B/(T_a+\theta)\} \tag{4.36b}$$

The analytical technique depends on the frequency range. For 'medium' and 'low' frequencies such that $\omega\tau$ is of the order of or less than unity a computer solution is required but at high frequencies, $\omega\tau \gg 1$, an approximate analysis is possible.

4.3.1.1. MEDIUM AND LOW FREQUENCY ANALYSIS

The instantaneous current in the circuit is

$$i(t) = \frac{E \sin \omega t}{S+R(T)}$$

and the instantaneous power dissipated in the thermistor is

$$P = R(T)\, i^2(t) = \frac{R(T)\, E^2 \sin^2 \omega t}{(S+R(T))^2} \tag{4.37}$$

Substitution of Equation 4.37 into Equation 4.1b gives

$$C\frac{d\theta}{dt} + K\theta = \frac{R(\theta)}{(S+R(\theta))^2} E^2 \sin^2 \omega t \tag{4.38}$$

Integration with respect to t brings the equation to a form suitable for analogue computation. Introduction of the identities

$$g(\theta) = \frac{R(\theta)}{(S+R(\theta))^2}$$

$$h(t) = E^2 \int_0^t \sin^2 \omega t \, dt$$

leads to

$$C\{\theta-\theta(0)\} = K\int_0^t \theta\, dt = \int_0^t g(\theta)\, dh(t) \tag{4.39}$$

Here $g(\theta)$ and $h(t)$ are known functions and furthermore $\theta(0) = 0$ if the e.m.f. is connected at $t = 0$ and the thermistor temperature at this instant is T_a. Equation 4.39 can readily be set up on a differential analyser and a solution for $\theta(t)$ found. Corresponding time-variations of R, i and v can then be calculated.

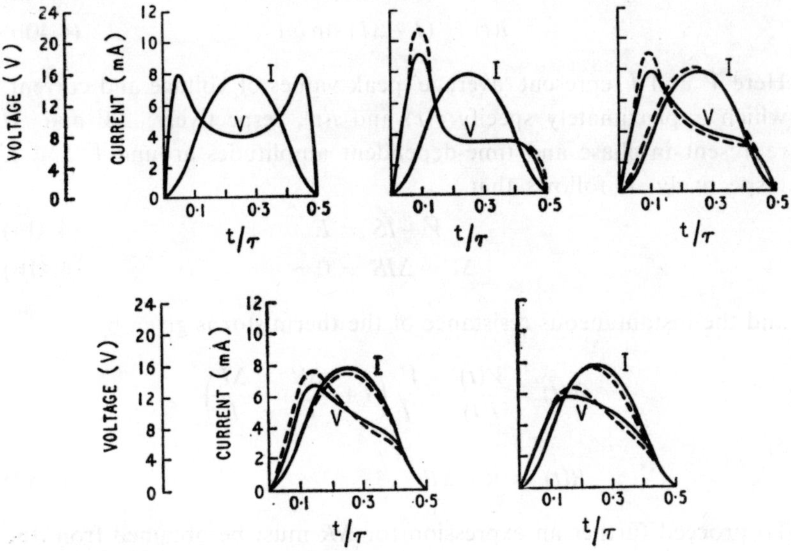

Fig. 4.7. Dependence of current and voltage on time (expressed as t/τ) for N.T.C. thermistor type A 1522/100 when fed from a sinusoidal voltage source of resistance 8 kΩ*; source frequency is (a) 0, (b) 0·1, (c) 0·2, (d) 0·4, (e) 0·8* Hz. *(after Ekelöf et al)*

The comparison between theory and experiment obtained by Ekelöf et al is shown in Fig. 4.7, for a Stantel thermistor type A1522/100 in series with a resistance S of 8 000 Ω with $E = 70$ V. The thermal time constant of this thermistor is $\simeq 0{\cdot}8$ s so that $1/\tau \approx 1{\cdot}3$ Hz. The differential analyser employed gave an accuracy of 0·25–0·5%. The ordinate axes in the figure indicate the steady state thermistor voltage, and current and the abscissae, time, normalised to τ, over one half cycle.

4.3.1.2. HIGH-FREQUENCY APPROXIMATION

If $\omega\tau > 1$ then the instantaneous resistance will vary very little over a period. For the same series circuit considered above it is then permissible to write the following expressions for the instantaneous voltage and current at the thermistor terminals.

$$v(t) = (\bar{V}+\Delta V)\sin\omega t \tag{4.40a}$$

$$i(t) = (\bar{I}+\Delta I)\sin\omega t \tag{4.40b}$$

Here $\bar{V}$ and $\bar{I}$ represent 'average' peak values of voltage and current, which approximately specify $v(t)$ and $i(t)$, respectively: ΔV and ΔI represent in-phase and time-dependent amplitudes around $\bar{V}$ and $\bar{I}$, respectively. It follows that

$$\bar{V}+\bar{I}S = E \tag{4.41a}$$

$$\Delta V+\Delta IS = 0 \tag{4.41b}$$

and the instantaneous resistance of the thermistor is given by

$$R(t) = \frac{V(t)}{I(t)} \approx \frac{\bar{V}}{\bar{I}}\left(1+\frac{\Delta V}{\bar{V}}-\frac{\Delta I}{\bar{I}}\right)$$

or

$$R(t) = \bar{R}+\Delta R \tag{4.42}$$

To proceed further an expression for ΔR must be obtained from the power-balance equation, 4.1b. This may be written

$$P-K\theta = C\frac{d\theta}{dt} = C\frac{d\theta}{dR}\cdot\frac{dR}{dT} \tag{4.43}$$

The relative variations in θ are much smaller than in $P = VI$, so we can put

$$P-K\theta \approx P-K\bar{\theta} = P-\bar{P} \tag{4.43a}$$

From Equation 4.43 an expression for dR can be obtained:

$$dR = \frac{(P-\bar{P})}{C}\frac{dR}{d\theta}dt \tag{4.44}$$

Since $dR/d\theta$ can be identified with $(\overline{dR/d\theta}) = \bar{F}\bar{R}K/\bar{P}$, in the present degree of approximation, this becomes

$$dR = \frac{(P-\bar{P})}{\bar{P}} \frac{\bar{F}\bar{R}}{\bar{\tau}} \tag{4.45}$$

where τ has also been replaced by its mean value $\bar{\tau}$.
Integration gives

$$R = \bar{R}\left\{1 - \frac{\bar{F}}{\bar{\tau}} \int_0^t \left(\frac{P}{\bar{P}} - 1\right) dt\right\} \tag{4.46}$$

Hence

$$\frac{\Delta R}{\bar{R}} = -\frac{\bar{F}}{\bar{\tau}} \int_0^t \left(\frac{P}{\bar{P}} - 1\right) dt \tag{4.47}$$

The applied power is given by

$$P = VI \approx \bar{V}\bar{I} \sin^2 \omega t$$

or

$$P = \bar{P}(1 - \cos 2\omega t) \tag{4.48}$$

where $\bar{P} = \bar{V}\bar{I}/2$. Substitution in Equation 4.47 and subsequent integration gives

$$\frac{\Delta R}{\bar{R}} \approx \frac{\bar{F}}{2\omega\bar{\tau}} \sin 2\omega t \tag{4.49}$$

Now from Equations 4.41 and 4.42 we can obtain

$$\frac{\Delta V}{\bar{V}} = \left(1 - \frac{\bar{V}}{E}\right) \frac{\Delta R}{\bar{R}} = \frac{S}{(S+\bar{R})} \frac{\Delta R}{\bar{R}} \tag{4.50a}$$

$$\frac{\Delta I}{\bar{I}} = -\frac{\bar{V}}{E} \frac{\Delta R}{R} = -\frac{\bar{R}}{(S+R)} \frac{\Delta R}{\bar{R}} \tag{4.50b}$$

Introducing ΔV and ΔI from Equation 4.50 into Equation 4.40 and paying regard to Equation 4.49, leads to the following final expressions for V and I:

$$V(t) = \bar{V}\left(\sin \omega t + \frac{D}{(1+\alpha)} (\cos \omega t - \cos 3\omega t)\right) \tag{4.51a}$$

$$I(t) = \bar{I}\left(\sin \omega t - \frac{D}{(1+\alpha)} (\cos \omega t - \cos 3\omega t)\right) \tag{4.51b}$$

where $\alpha = \bar{R}/S$ and $D = \bar{F}/4\omega\tau$.

The current lags the voltage at the fundamental frequency as expected for the N.T.C. thermistor, which we have already seen has inductive properties for small signals. The magnitude of the phase angle is

$$\phi = \tan^{-1} \frac{D}{1 - \dfrac{D^2 \alpha}{(1+\alpha)^2}} \tag{4.52}$$

or $\phi \simeq D$ if $D \ll 1$.

If $S = \infty$, i.e. $\alpha = 0$, only the voltage is distorted:

$$V = \bar{V}(\sin \omega t + D(\cos \omega t - \cos 3\omega t)) \tag{4.53}$$

Correspondingly if $S = 0$, i.e. $\alpha = \infty$ only the current is distorted:

$$I = \bar{I}(\sin \omega t - D(\cos \omega t - \cos 3\omega t)) \tag{4.54}$$

4.3.2. Transient Analysis

It is convenient to use a semi-graphical approach.[5] The same series circuit as analysed above will be considered here for the case in which, at $t = 0$, the circuit e.m.f. is changed from E_1 to E_2 and the series resistance from S_1 to S_2. These conditions are shown by the corresponding load lines *I* and *II* drawn on the static characteristic for an N.T.C. thermistor at ambient temperature T_a, in Fig. 4.8

For $t < 0$ the circuit is in the stationary state, represented by point 1 (V_1, I_1), which is the intersection between the thermistor $V-I$ characteristic and load line I, determined by the equation

$$V + S_1 I = E_1 \tag{4.55}$$

The thermistor resistance $R_1 = V_1/I_1$ is on the isothermal (linear) characteristic from the origin, which passes through point 1. At $t = 0$ the change in E and S to the new values E_2 and S_2 force the dynamic point V, I to move from load line *I* to load line *II*. Because of the thermal inertia of the thermistor, abrupt changes of thermistor resistance are not possible, so at $t = 0$ the operating point is constrained to move along the isothermal characteristic passing through point 1 to the corresponding point 1′ on load line *II*. Hence the thermistor voltage and current change instantaneously from V_1, I_1 to V_1', I_1'.

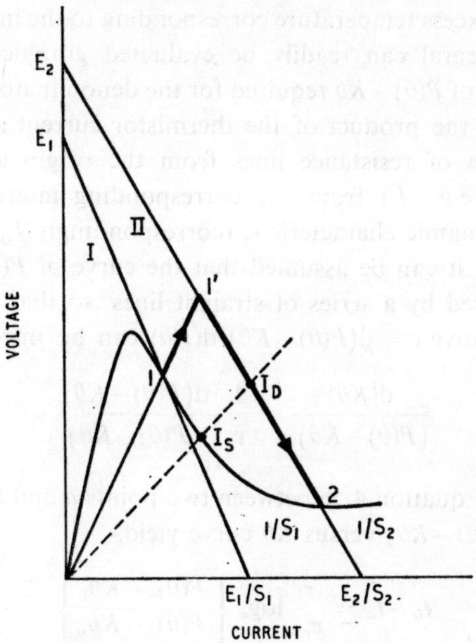

Fig. 4.8. Static and dynamic characteristics of an N.T.C. thermistor for transient analysis

At the same time the power supplied increases from V_1I_1 to $V_1'I_1'$, while the rate of heat loss is unchanged. Hence for $t > 0$ the thermistor temperature increases and its static resistance decreases. Therefore the dynamic operating point moves downwards on load line *II* from 1′ until it reaches point 2, its intersection with the static characteristic, which represents the final stationary state.

The dynamic characteristic is therefore represented by 1–1′–2. Since instantaneous changes in *E* and *S* have been assumed it is only necessary to determine the movement of the dynamic point from 1′ to 2. This is governed by the time-dependence of the thermistor excess temperature θ, or the corresponding inverse function, which can be written, by integration of Equation 4.1b;

$$t = \int_{\theta_1}^{\theta} \frac{\tau(\theta)\,\mathrm{d}(K\theta)}{(P(\theta)-K\theta)} \tag{4.56}$$

Here θ_1 is the excess temperature corresponding to the initial operating point. The integral can readily be evaluated graphically if $\tau(\theta)$ is known. Values of $P(\theta)-K\theta$ required for the denominator are obtained by subtracting the product of the thermistor current and voltage at the intersection of resistance lines from the origin and the static characteristic (e.g. I_s) from the corresponding intersection of this line and the dynamic characteristic (correspondingly I_D).

Alternatively it can be assumed that the curve of $P(\theta)-K\theta$ versus $K\theta$ is represented by a series of straight lines, so that the piecewise-constant derivative $c = \mathrm{d}(P(\theta)-K\theta)/\mathrm{d}(K\theta)$ can be introduced. Then

$$\frac{\mathrm{d}(K\theta)}{(P(\theta)-K\theta)} = \frac{1}{c}\,\frac{\mathrm{d}(P(\theta)-K\theta)}{(P(\theta)-K\theta)} \tag{4.57}$$

Integration of Equation 4.56 between two points a and b on a straight part of the $(P(\theta)-K\theta)$ versus $K\theta$ curve yields

$$t_b - t_a = \frac{\tau}{c_{ab}} \log_e \left\{ \frac{P(\theta)_b - K\theta_b}{P(\theta)_a - K\theta_a} \right\} \tag{4.58}$$

If, in particular, the internal power dissipation is small, $c = -1$ and Equation 4.58 becomes

$$t_b - t_a = -\tau \log_e \frac{\theta_b}{\theta_a}$$

or

$$\theta_b = \theta_a \exp\{-(t_b - t_a)/\tau\} \tag{4.59}$$

4.3.3. Practical Considerations

Experimental work has been carried out for comparison with the above transient theory.[5] In Fig. 4.9a the variation of static resistance with time is shown for a bead-type N.T.C. thermistor, type A1522/100, when the initial heating was negligible (I_1 of the order of μA), with $E = 20$ V and $I_2 = 3$ and 8 mA. Part (b) shows the corresponding change when switching in the opposite sense: i.e. from a large to a small current. A graphical integration of Equation 4.56 was used to obtain

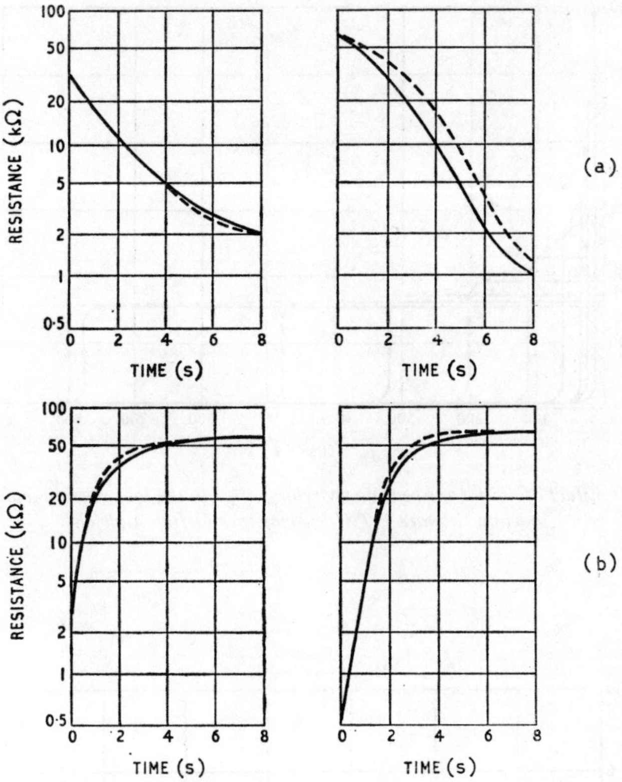

Fig. 4.9. Transient characteristics of thermistor type A 1522/100: (a) heating; final current (i) 3 mA, *(ii) 8* mA. *(b) cooling; initial current (i) 3* mA, *(ii) 8* mA. *(after Ekelöf et al)*

the theoretical values for $R(t)$. The integration was based on measured values of the static characteristic and of τ.

The agreement between theory and experiment is good only for sudden decreases of current or small increases of current, i.e. for cooling or slow-heating. For fast heating considerable deviations occur, especially in the earlier parts of the transients. The explanation is that the temperature distribution within the thermistor is not homogeneous and R is likely to be different under steady-state and transient conditions, so that the use of static values of $R(\theta)$ can be misleading.

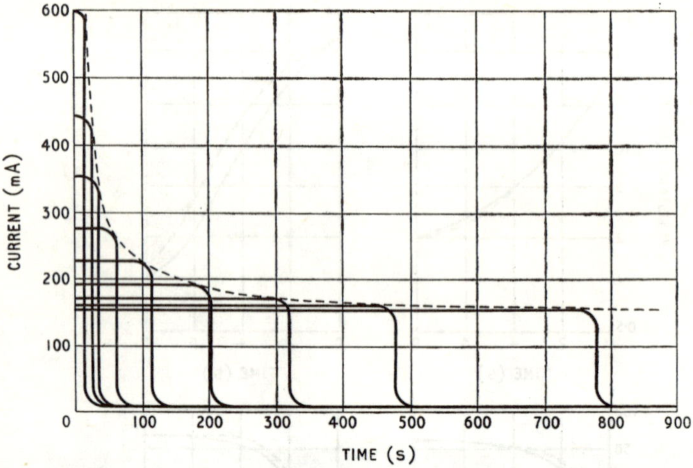

Fig. 4.10. Effect of initial current on switching time to the low current condition for barium titanate P.T.C. thermistors (After Andrich)

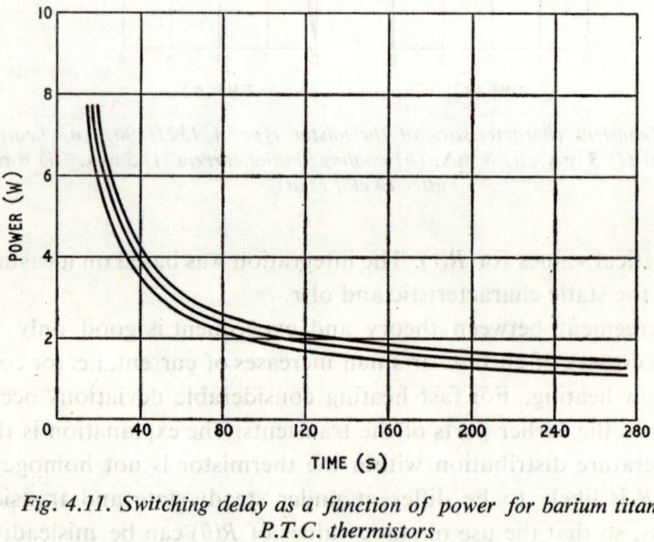

Fig. 4.11. Switching delay as a function of power for barium titanate P.T.C. thermistors

Experiments[5] on rod-type N.T.C. thermistors also gave results which were in good agreement with theory over a wide range of conditions.

Andrich[8] has given data on transient conditions for P.T.C. thermistors. In Fig. 4.10 typical curves show the decay of current from a high to low value in passing from a low to high voltage operating point as a result of either increasing applied e.m.f. or temperature, or decreasing the series resistance: the initial bias current is the parameter. The envelope of the curves is approximately hyperbolic, with asymptotes $\tau = 0$ and $I = I_{max}$, the latter being the current corresponding to the peak of the current-voltage characteristic. The information can also be presented with dissipated power as the ordinate as shown in Fig. 4.11. The general equation for these curves is

$$(P - P_0)\tau = \text{constant}$$

where P is the initial power and P_0 the power necessary to bring the thermistor up to the switching temperature T_s (see Fig. 2.13). The dependence of switching time on ambient temperature is included in the following expression

$$\tau = \frac{C(T_s - T_a)}{P - K(T_s - T_a)} \tag{4.60}$$

4.4. FLUCTUATIONS IN THERMISTORS

Thermal equilibrium is a dynamic, not a static condition. In any solid matter in thermal equilibrium there is a spontaneous fluctuation of the total energy contained in the fixed atoms or molecules and the free electrical charges. This is manifest as a fluctuation in the temperature of the material. There are also fluctuations in the instantaneous densities and spatial distribution of the free charges. The overall fluctuations associated with the free charges give rise to associated terminal electrical fluctuations of voltage and current, which are generically referred to as 'noise', since fluctuations of this kind were first studied in detail in relation to the output fluctuations of high-gain audio amplifiers.

Below, brief consideration is given to the electrical and thermal fluctuations in thermistors, which are of significance in low-level applications such as the detection and measurement of radiation, which is of the subject matter of Chapter 6. The magnitude of such fluctuations imposes a limit on the smallest signal which can be detected.

4.4.1. Electrical Fluctuations[9, 10]

Thermistors are essentially semiconductors and for this general class of materials there is now ample evidence that the electrical noise depends in a complicated way on current, voltage, temperature and frequency, but that there are certain well-defined contributions to the total noise, which vary only in degree from one material to another and from one device to another. The most common way of specifying basic noise phenomena is in terms of the spectral density of the mean square value of the noise quantity concerned. For example, if the total mean square value of the noise voltage associated with a particular device or process is $\overline{|\Delta V|^2}$, then the spectral density function of mean square noise voltage is defined as $S_v(f)$, with

$$\overline{|\Delta V|^2} = \int_0^\infty S_v(f)\,\mathrm{d}f$$

i.e. $S_v(f)$ represents the contribution to the total mean square noise voltage contained in unit frequency band at frequency f. $S_v(f)\,\mathrm{d}f$ is the corresponding contribution in a band of frequencies of width $\mathrm{d}f$ at f.

A general form for the spectral density of the mean square noise voltage of a semiconductor of resistance R carrying a current I is as follows

$$S_v(f) = \left[\frac{K_1 I^\alpha R^2}{f^\beta} + \frac{K_2 I^\gamma}{(1+\omega^2\tau_0^2)} + 4kTR\right]\mathrm{d}f \qquad (4.61)$$

where K_1, K_2, α, β, and γ are specific constants, T is the absolute temperature, τ_0 a characteristic relaxation time or lifetime and k Boltzmann's constant. The first component in Equation 4.61 is 'flicker' or '$1/f$' noise, so-called because its spectral density falls approximately inversely as the frequency. Flicker noise originates at the end contacts, on the surface and at grain boundaries and dislocations. It is sensitive to the method of manufacture, but is much less intense than the similar noise component associated with carbon resistors of comparable resistance values. Both van Vliet et al and Brophy found α to be about 1·2–1·3, for N.T.C. types.

Various mechanisms can give rise to the second type of component in Equation 4.61. One which has been fully investigated in single-crystal semiconductors is 'generation-recombination' noise. It arises because of fluctuations in the instantaneous rates of generation and recombination of carriers, which cause corresponding fluctuations in the instantaneous free carrier densities. Another cause of this particular spectral form is trapping of free carriers.

The final term in Equation 4.61 represents the ubiquitous thermal or Johnson noise associated with the small-signal resistance R.

Little published information is available on electrical noise in thermistors. N.T.C. types have been investigated in some detail by van Vliet et al and Brophy. They find that the '$1/f$' noise component is very much less than that in carbon resistors of comparable value. Typical 'excess' noise spectra due to van Vliet are shown in Fig. 4.12: these continuous curves do not include the thermal noise which is represented separately by the broken curves. The overriding '$1/f$' type of noise is evident at frequencies below 100 Hz. Above that noise associated with a single controlling time constant is evident at low currents. Brophy did not observe this component in flake bolometers. Both van Vliet and Brophy found α as defined in Equation 4.61 to lie between 1·2 and 1·3 and β to be slightly greater than unity.

Van Vliet found that for the single time-constant noise component for a rod thermistor the corresponding spectral density of noise current to voltage (eliminating the temperature dependence of R) was approximately proportional to I and then becomes independent of current beyond 200 μA. He considered possible mechanisms for the generation of this noise. On the assumption that it is true $g-r$ noise the corner frequency suggests a lifetime τ_0 of $1{\cdot}6\times10^{-3}$ s, leading to a value for transit time of $1{\cdot}6\times10^{-1}$ s. The value of carrier mobility calculated using this is 2×10^{-4} cm^2/Vs, which is much too low.

Other possible mechanisms considered by van Vliet are the fluctuation in heat transfer to the surroundings or fluctuations in heat conductance caused by fluctuating heat diffusion. The first of these possibilities is incompatible with the thermal time constant τ of the thermistor which

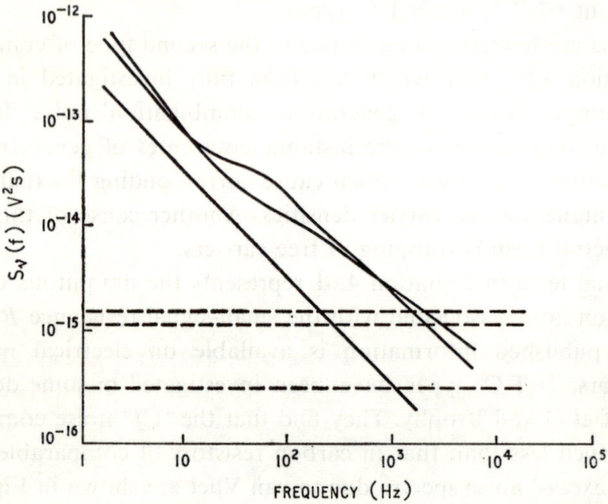

Fig. 4.12. Dependence of spectral density of noise voltage of a rod-type N.T.C. thermistor on frequency. Current is the parameter; (a) $I = 600$ μA, $R = 20$ kΩ; (b) $I = 167$ μA, $R = 80$ kΩ; (c) $I = 111$ μA, $R = 100$ kΩ. (after van Vliet et al)

was about 1 s. The second thermal mechanism leads to a time constant $\tau_0 = C\,dl^2/4\lambda$ where C is the thermal capacity, d the density, l the length of the sample and λ the heat conductivity; this was found to be compatible with the observed noise time constant, but an $f^{-3/2}$ fall-off is predicted at high frequencies and this was not investigated in detail experimentally.

There is little information on noise in P.T.C. thermistors. For single-crystal Si types the electrical noise is, however, well covered by the basic work done on silicon itself. For barium titanate types Saburi and Wakino[12] report that the noise behaviour between 300 and 3 000 Hz is similar to that of carbon resistors.

4.4.2. Thermal Fluctuations[11]

As outlined at the beginning of Section 4.4 the temperature of a body is determined by statistical interchange of energy between it and its surroundings and in consequence the temperature fluctuates in a random manner about a mean value. The expression for the spectral density of the mean square temperature fluctuation in a bandwidth df at a frequency $f = \omega/2\pi$ is of somewhat similar form to the Johnson noise term in Equation 4.61, but is now frequency-dependent, since the controlling thermal time constant for temperature fluctuations is many orders of magnitude greater than that which governs Johnson noise—the mean free time of free carriers between collisions:

$$\overline{|\Delta T(f)|^2} = \frac{4kT^2K}{K^2+\omega^2C^2}\,\mathrm{d}f \tag{4.62}$$

Here k is again Boltzmann's constant, K is the thermal conductance nd C the thermal capacitance of the thermistor. This expression is alid even if K and C depend on frequency.

A corresponding equation gives the spectral distribution of the mean square power flow, $\overline{|\Delta P_r|^2}$, to the thermistor:

$$\overline{|\Delta P_r|^2} = 4kT^2K\,\mathrm{d}f \tag{4.63}$$

Use of this equation will be made in Section 6.2.4 in connection with the ultimate limits of sensitivity of the thermistor bolometer.

REFERENCES

1. EKELÖF, S. and KIHLBERG, G. 'Theory of the thermistor as an electric circuit element', *Chalmers tek. Högsk. Handl.*, (1954) *Avd Elektroteknik* 40, Nr. 42
2. BURGESS, R. E. 'The a.c. admittance of temperature-dependent circuit elements', *Proc. Phys. Soc.*, **68,** 766 (1955)
3. HYDE, F. J. 'Impedance of a thermistor at low frequencies', *J. Electron.*, **1,** 303 (1955)
4. SAPOFF, M. and OPPENHEIM, R. M. 'Theory and applications of self-heated thermistors', *Proc. Instn elect. electronic. Engrs.*, **51,** 1292 (1963)
5. EKELÖF, S., BJÖRK, N. and DAVIDSON, R. 'Large-signal behaviour of directly

heated thermistors', *Chalmers tek. Högsk. Handl.*, (1957) *Avd Elektroteknik*, **56,** Nr. 185

6. CANDY, C. J. N. 'The specification of the properties of the thermistor as a circuit element in very low frequency systems', *Proc. Instn. elect. Engrs*, **103B,** 398 (1954)
7. BJÖRK, N. and DAVIDSON, R. 'Small-signal behaviour of directly-heated thermistors', *Chalmers tek. Högsk. Handl.*, (1955) *Avd Elektroteknik*, **48,** Nr. 169
8. ANDRICH, E. 'Properties and applications of P.T.C. thermistors', *Electron. Appl.* (Neth.), **26,** 123 (1966)
9. VAN VLIET, K. M., VAN LEEUWEN, C. K., BLOCK, J., and RIS, C., 'Measurements on current noise in carbon resistors and in thermistors', *Physica*, **20,** 481 (1954)
10. BROPHY, J. J. 'Current noise in thermistor bolometer flakes', *J. Appl. Phys.*, **25,** 222 (1954)
11. SMITH, R. A., JONES, F. E., and CHASMAR, R. P. *The detection and measurement of infrared radiation* Clarendon Press, 204 (1957)

CHAPTER 5

Small-signal A.C. Applications

5.1. INTRODUCTION

The theoretical and practical considerations of the preceding chapter have drawn attention to the small-signal properties of thermistors. These are represented by the equivalent circuits of Fig. 4.3.

An important application at frequencies much greater than the reciprocal of the thermal time constant is in providing a continuously variable resistance. At low frequencies, such that the thermal inertia is manifest as an equivalent inductance (N.T.C. types) or an equivalent capacitance (P.T.C. types) there is potential application in phase-shifting, differentiating and integrating circuits.

Such application will be outlined in this chapter.

5.2. THE THERMISTOR AS A VARIABLE RADIO FREQUENCY RESISTOR[1–4]

In r.f. circuits the need often arises for a continuously-variable resistance. Problems arise with wirewound resistances because of the relatively large self-inductance and stray capacitance. Although variable carbon resistors can be made with relatively small inductances the variability of contact resistance precludes their use for precise work.

For all except the smallest values of resistance the useful frequency range is limited to a few tens of kHz.

For $\omega\tau \gg 1$ the resistance of a thermistor is R_∞, which is also its d.c. resistance. It is therefore possible to incorporate a thermistor in a d.c. bridge circuit, which can be balanced to obtain a given value

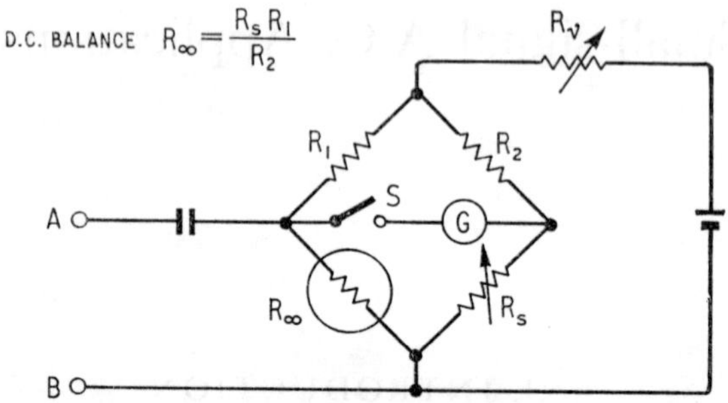

Fig. 5.1. Basic bridge circuit incorporating a thermistor to generate a variable radio frequency resistor

of R_∞, in terms of precision d.c. components, and simultaneously connect it, via a blocking capacitor, into an a.c. circuit. The principle is illustrated in Fig. 5.1. A d.c. balance is obtained where R_s is set to satisfy the demanded value of R_∞, by adjusting R_v; the value of R_∞ is given by

$$R_\infty = \frac{R_s R_1}{R_2} \tag{5.1}$$

If the switch S is opened after the d.c. balance has been obtained, then the equivalent a.c. resistance in parallel with R_∞ is R_1 plus the parallel combination of R_v and R_2 and R_s in series. If R_1 and R_2 are chosen to be $\gg R_\infty$ then this effective combination can itself be made large relative to R_∞, thereby reducing the associated error in assumming that the a.c. resistance between A and B is R_∞.

Soldi[1] has described a self-balancing bridge incorporating a bead-type directly-heated N.T.C. thermistor in which R_v was replaced by a constant-current pentode and the current through this and hence the value of R_∞ were determined by the out-of-balance voltage of the bridge. Operation was in the frequency range 1–10 MHz, with values of R_∞ between 1 and 50 kΩ. A setting accuracy of better than 2% was obtained up to a frequency-resistance product of 50 MHz kΩ. At higher values than this accuracy fell off due to the effect of lead inductance and stray capacitance.

For directly-heated thermistors the effects due to stray reactances associated with series inductance and distributed capacitance have been considered by Walker[2]. He found that up to 400 Hz variability of R_∞ due to skin effect is negligible and series inductance did not have a significant effect. However, he found that distributed capacitance had a considerable influence on the performance of thermistors having a resistance of more than a few thousand ohms at u.h.f.

Hyde and Price[3] used a modified version of the bridge shown in Fig. 5.1, which did not require large values of R_1 and R_2, for evaluating transistor parameters. They incorporated a radio-frequency choke in series with R_1 and took account of its d.c. resistance in determining the value of R_∞ from the d.c. balance equation for the bridge. Whitehead[4] has described a programme-fading circuit in which a thermistor is used as a remotely-controlled variable resistor.

5.3. THE THERMISTOR AS A LOW FREQUENCY CIRCUIT ELEMENT[5, 6]

5.3.1. Phase-advancing (Differentiating) Circuit

The use of an N.T.C. thermistor will be considered for the simple circuit shown in Fig. 5.2a and in equivalent form in 5.2b. The d.c. bias circuit is not shown. The voltage transmission is given by

$$T_d = \frac{V_2}{V_1} = \frac{R_1+R_0+j\omega\tau(R_1+R_\infty)(R_\infty+R_0)/2R_\infty}{R+R_1+R_0+j\omega\tau(R+R_1+R_\infty)(R_\infty+R_0)/2R_\infty} \qquad (5.2)$$

The locus of the tip of the corresponding vector is a semicircle as shown in Fig. 5.3. Two conditions are portrayed. On the left the d.c. bias is such that R_1+R_0 is positive: the phase shift is therefore less

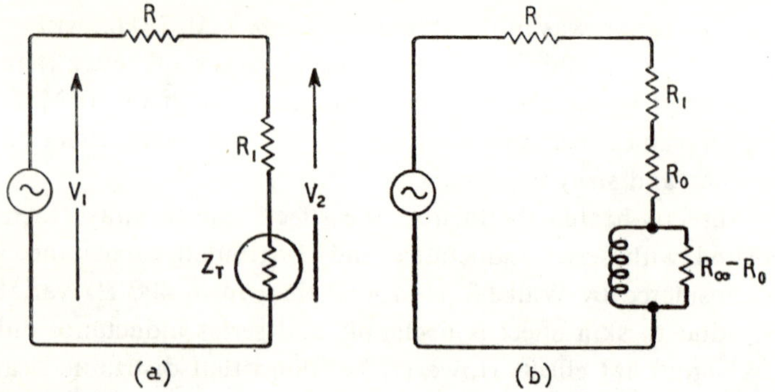

Fig. 5.2. (a) Phase advancing circuit incorporating N.T.C. thermistor; (b) equivalent circuit

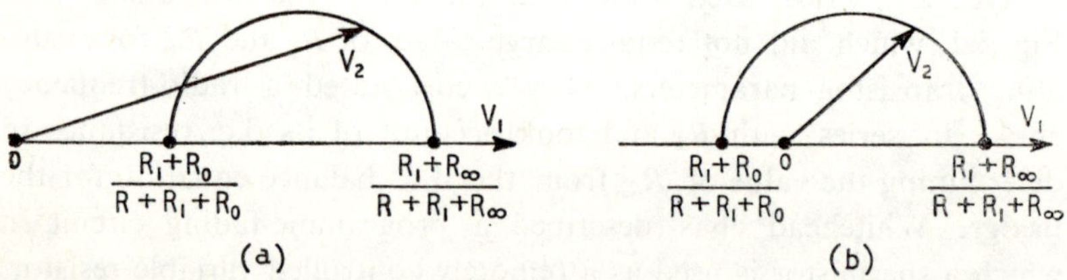

Fig. 5.3. Phase advance transmission locus; (a) $R_1+R_0 > 0$, (b) $R_1+R_0 < 0$

than 90° at all frequencies. On the right the thermistor is biased sufficiently far beyond turnover that R_1+R_0 is negative: phase advance up to 180° can be achieved in this case.

The circuit time constant is (see Section 4.2.3)

$$\tau_c = \tau \frac{(R_\infty+R_0)(R+R_1+R_\infty)}{2R_\infty(R+R_1+R_0)} \tag{5.3}$$

A choice of additional series and shunt resistances enables the circuit time constant to be adjusted to any desired value. For the circuit configuration shown in Fig. 5.4 it is

$$\tau_c = \tau \frac{(R_\infty+R_0)(R_s+R_\infty)}{2R_\infty(R_s+R_0)} \tag{5.4}$$

where R_s is the resistance presented to the thermistor:

$$R_s = R_2 + R_3(R+R_1)/(R_3+R_1+R)$$

If the circuit is to be stable R_s must be greater than R_0 and for the time constant to be large $R_s \approx R_0$. Hence the larger is the demanded time constant the greater is the possibility of instability.

To illustrate a possible design procedure it will be assumed that the thermistor impedance is specified and values of R, R_1, R_2 and R_3

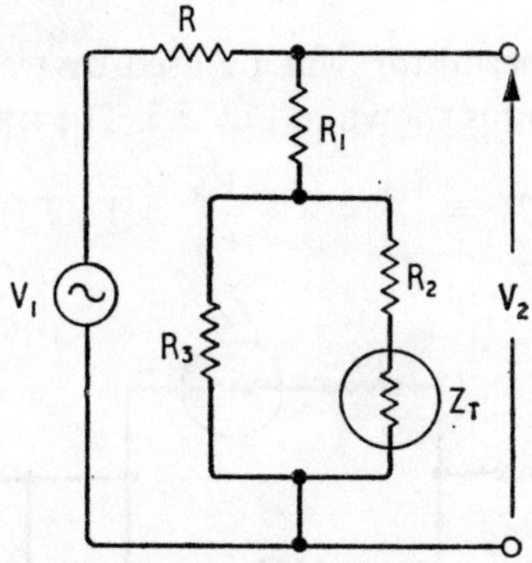

Fig. 5.4. General phase-advancing network, incorporating N.T.C. thermistor

have to be chosen to define a required ratio of high to low frequency transmission $T_{dr} = T_{d\infty}/T_{d0}$, and a circuit time constant $\tau_c = x\tau_e$ where τ_e is the 'effective thermal time constant' of the thermistor, defined in Equation 4.23. Hence

$$x\tau_e = \left(\frac{R_s+R_\infty}{R_s+R_0}\right)\tau_e \tag{5.5}$$

and

$$R_s = \frac{R_\infty - xR_0}{x-1} \tag{5.6}$$

Since R_0 and R_∞ have been prescribed, R_s is fixed. The transmission ratio is given by

$$T_{dr} = \frac{(R_y+R_\infty)(R_s+R_0)}{(R_y+R_0)(R_s+R_\infty)} = \left(\frac{R_y+R_\infty}{R_y+R_0}\right)\frac{1}{x} \tag{5.7}$$

where

$$R_y = R_2 + R_1R_3/(R_1+R_3) \tag{5.8}$$

is therefore also fixed. R should be made as small as possible, so that the attenuation will be small. The minimum value is that of the voltage source. Either R_1, R_2 or R_3 may be given an arbitrary value, say zero, and the others calculated using Equations 5.6 and 5.8.

5.3.2. Phase-retarding (Integrating) Circuit

Again an N.T.C. thermistor will be used by way of example. The general form of circuit is shown in Fig. 5.5. The transmission is

$$T_i = \frac{V_2}{V_1} = 1 - \frac{V_3}{V_1} = 1 - T_d \tag{5.9}$$

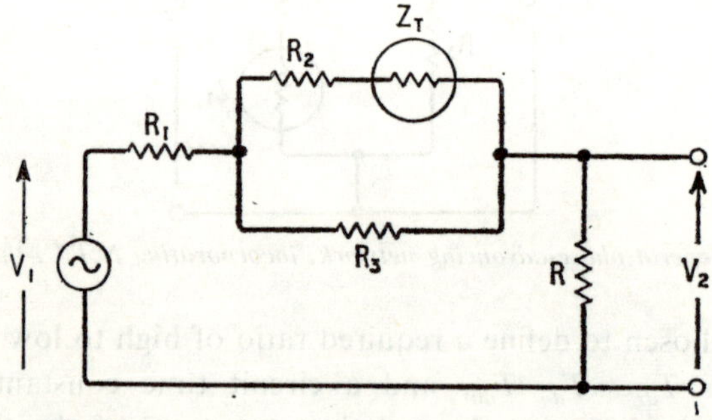

Fig. 5.5. General phase-retarding (integrating) network, incorporating N.T.C. thermistor

It follows that the circuit time constant is the same as that of the differentiating network considered above. In this case the resistance R_s presented to the thermistor is given by

$$R_s = R_2 + \frac{R_3(R_1+R)}{R+R_1+R_3} = \frac{R_\infty - xR_0}{x-1} \tag{5.10}$$

The ratio T_{ir} of the high to low frequency transmissions is

$$T_{ir} = \frac{1-T_{di}}{1-T_{d0}} = \frac{1-\dfrac{(R_y+R_\infty)(R_3+R_1)}{(R_s+R_\infty)(R_3+R_1+R)}}{1-\dfrac{(R_y+R_0)(R_3+R_1)}{(R_s+R_0)(R_3+R_1+R)}}$$

$$= \frac{1}{x}\,\frac{(R_2+R_3+R_\infty)}{(R_2+R_3+R_0)} \tag{5.11}$$

From this

$$R_2+R_3 = \frac{R_\infty - xT_{ir}R_0}{xT_{ir}-1} \tag{5.12}$$

Equations 5.6 and 5.12 may be used to design the integrating network.

Practical forms of biasing circuits for N.T.C. thermistors in phase-advancing and retarding circuits, have been given by Candy, who also suggested the above applications.

5.3.3. Thermistor Oscillators[6–8]

The small-signal equivalent circuits of Fig. 4.3 show that over a range of frequencies both N.T.C. and some P.T.C. thermistors biased beyond turnover can have a terminal admittance represented by a parallel combination of negative conductance and positive and negative susceptance, respectively. In principle therefore oscillation can be induced in each case by terminating the thermistor by an equal susceptance of opposite sign. In fact, since the N.T.C. thermistor has an equivalent terminal inductance and the P.T.C. thermistor a corresponding capacitance it is possible to devise a combination of N.T.C. and P.T.C. thermistors, exhibiting resonance, so that electrical and thermal oscillations occur. For simpler situations involving only one thermistor a passive, external inductor or capacitor can provide the necessary resonating reactance or susceptance.

In practice it is evident that oscillations in thermistor circuits follow a relaxation rather than simple harmonic pattern, but for a first-order appreciation a small-signal analysis is useful since it gives the appropriate initial conditions. For example it follows from the equivalent circuit for an N.T.C. thermistor biased beyond turnover that the terminal conductances will be negative for all angular frequencies less than

$$\omega_{\max} = \frac{2(-R_0R_\infty)^{1/2}}{\tau(R_\infty+R_0)} \tag{5.13}$$

This is the theoretical maximum frequency at which the thermistor can be made to oscillate when shunted by a suitable lossless capacitor.

In practice the actual upper frequency of oscillation will be somewhat less because the resistance R of the bias circuit cannot be made infinite. When this is included it can be shown that oscillation at $\omega\ (< \omega_{max})$ can occur if

$$R \geqslant -R_\infty \left\{ \frac{(R_0/R_\infty)\omega_{max}^2 - \omega^2}{\omega_{max}^2 - \omega^2} \right\} \tag{5.14}$$

Provided that this condition obtains the value of C required to support oscillation at ω is

$$C = \frac{(2\tau/R_\infty)\,(R_\infty - R_0)/(R_\infty + R_0)}{\omega^2\tau^2 + \{4R_0^2/(R_\infty + R_0)^2\}} \tag{5.15}$$

REFERENCES

1. SOLDI, M. 'Applications of thermistors to bridges for measuring low conductance at high frequencies', *Alta Freq.*, **21,** 243 (1952)
2. WALKER, J. 'Thermistors at high frequencies' *Wireless Eng.*, **24,** 28 (1947)
3. HYDE, F. J. and PRICE, T. E. 'Measurement of the high-frequency base resistance and collector capacitance of drift transistors', *J. Elect. Control.* 1959, **6,** p. 347.
4. WHITEHEAD, R. C. 'A Programme-fading circuit' *B.B.C. Quarterly*, **8,** No. 4 (1953/4)
5. CANDY, C. J. N. The specification of the properties of the thermistor as a circuit element in very low frequency systems' *Proc. Instn. elect. Engrs.*, **103B,** 398 (1955)
6. HYDE, F. J. 'Impedance of a thermistor at low frequencies', *J. Electron.* **1,** 303 (1955)

CHAPTER 6

Detection and Measurement of Radiation

6.1. INTRODUCTION

There are several ways in which electromagnetic radiation can be absorbed by thermistor materials, the phenomena involved being dependent on the wavelength of the radiation. If this is short enough then free carriers can be created by ionisation of host atoms or of impurity atoms (photoconductivity). In the infrared region of the spectrum (wavelengths between approximately 1 and 1000 μ) the absorption involves inter and intra-atomic excitations. At longer wavelengths, including the microwave region, the absorption is primarily via the free charges, which take energy from the field and share it with the lattice as a result of collision processes. The absorption of radiation results in a rise in temperature of the material and so a change in resistance. The same is true when a beam of particles is incident on the material: in this case the kinetic energy of the beam is converted into heat.

The rise in temperature and change in resistance of suitably-designed thermistor elements permit their use for detection and, in some cases, measurement of radiation. A typical family of characteristics for an N.T.C. thermistor with absorbed power as the parameter, is shown in Fig. 1.4.

Below, we consider in turn, infrared detection, microwave detection and measurement, and finally, the detection and measurement of the energy of particle beams.

6.2. INFRARED DETECTION

6.2.1. Thermal Detectors

As a result of developments in detectors and optical materials, three natural, though somewhat arbitrary divisions, of the infrared spectrum, are commonly used. These are:

(a) the near infrared; $\lambda \approx 0{\cdot}7\text{–}1{\cdot}5\mu$
(b) the intermediate or middle infrared; $\lambda \approx 1{\cdot}5\text{–}5\text{–}6\mu$
(c) the far infrared; $\lambda \approx 5\text{–}6\text{–}1\,000\mu$

The region in which thermistor bolometers are most useful is the 'far infrared', in which the absorbed energy is converted into heat. They therefore belong to the class of 'thermal detectors', which also includes thermocouples, thermopiles and metal bolometers (barretters) and superconducting detectors, among others.[1–3] Each type has advantages which makes it useful for special applications. For thermistor bolometer detectors these include use[4] in radiometers, pyrometers and other infrared instruments with non-destructive applications to industry, such as automatic product-analysis, spectroscopy and the automatic control of manufacturing operations. They also find medical use in the infrared thermograph;[5] this records the temperature-distribution of the body, a knowledge of which is useful in the diagnosis of some diaseases.

6.2.2. Figures of Merit for Infrared Detectors

The important questions to be asked about an infrared detector are

a. What is the smallest radiant power that can be detected?
b. What signal is generated per unit radiant incident power?

*c. What modulation frequencies can be used?
d. In what way does the signal depend on the wavelength of the radiation?

The information asked for in the last question is usually given in graphical form. The first three questions are covered by the quantities defined below.[1–3, 6]

6.2.2.1. NOISE EQUIVALENT POWER (N.E.P.); P_n

This is the value of the sinusoidally-modulated radiant power falling on a detector, which will give rise to an r.m.s. signal voltage equal to the r.m.s. noise voltage from the detector. The temperature of the blackbody radiation source is usually 500 K. The reference bandwidth in which the noise is generated is usually either 1 or 5 Hz and the centre frequency, which is the modulation frequency, is usually in the audio or sub-audio range. For thermistor bolometers the reference temperature is about 295 K and the reference area is usually 1 cm². In quoting N.E.P. all of the above should be specified.

An experimental determination of N.E.P. involves the determination of r.m.s. signal/noise voltage ratio $\overline{|V_n|^2}^{1/2}$ for a bandwith Δf, for a known amount of radiant power, P_r, and then, by proportion, finding P_n as follows;

$$P_n = P_r \frac{\overline{|V_n|^2}^{1/2}}{V_s} \cdot \frac{1}{(\Delta f)^{1/2}} \tag{6.1}$$

The above relationship arises since the total mean square noise voltage in the bandwith Δf is $\overline{|V_n|^2}$, so that the mean square noise voltage per unit bandwidth $\overline{|V_n|^2}/\Delta f$ and the r. m. s. noise voltage is correspondingly $\overline{|V_n|^2}^{1/2}/(\Delta f)^{1/2}$. P_n is specified in W Hz$^{-1/2}$, although the Hz$^{-1/2}$ is commonly dropped.

* This question arises because it is common to use an infrared detector in an a.c. mode by 'chopping' (modulating) the radiation at some low frequency. This permits a.c. amplification of the signal generated, with the attendant advantages over d.c. amplification. The chopping frequency must be sufficiently low that the detector is able to follow the excitation.

6.2.2.2. DETECTIVITY, D

This is simply the reciprocal of N.E.P. For a good detector D is large

6.2.2.3. NOISE EQUIVALENT INPUT (N.E.I.)

This is the radiant power per unit area of the detector required to give a signal/noise ratio of unity. N.E.I. is therefore N.E.P. divided by the detector area A:

$$\text{N.E.I.} = \frac{P_n}{A} = \frac{1}{AD} \tag{6.2}$$

The commonly-used units of N.E.I. are W cm^{-2} $\text{Hz}^{-1/2}$

6.2.2.4. 'DEE-STAR', D*

Since most infrared detectors, including thermistor bolometers, have an N.E.I. directly or approximately proportional to the square root of the detector area, an 'area-independent' figure of merit is in general use. This is D^* and it is obtained by multiplying D by the square root of th e area:

$$D^* = DA^{1/2} = A^{1/2}/P_n \tag{6.3}$$

The units are cm $\text{Hz}^{1/2}$ W^{-1}. Current usage is to refer to D^* as the detectivity, D now being rarely used.†

6.2.2.5. RESPONSIVITY

This is defined as the r.m.s. signal voltage generated per unit radiant power, P_r:

$$\mathcal{R} = \frac{V_s}{P_r} \tag{6.4}$$

† If, as is usual, the wavelength λ is specified, each of the quantities defined carries a suffix λ; e.g. D^*_λ.

The temperature of the blackbody generating P_r is usually 500 K. The units of $\mathcal{R}$ are VW^{-1}. It follows from Equations 6.1, 6.3 and 6.4 that

$$\mathcal{R} = \frac{\overline{|V_n|^2}^{1/2}}{P_n(\Delta f)^{1/2}} = \frac{\overline{|V_n|^2}^{1/2} D^*}{(A\,\Delta f)^{1/2}} \tag{6.5}$$

6.2.2.6. DETECTOR TIME CONSTANT

This is the 'response time' of the detector; i.e. the time after the application of a step-function stimulus for the controlled parameter to change by $1/e$ of its total change. For thermistor bolometers the appropriate time constant is the effective circuit time constant τ_c defined in Equation 4.22.

6.2.3. Practical Thermistor Bolometers

A typical thermistor bolometer element designed for bridge working is shown in Fig. 6.1a. Such designs incorporate two extremely thin 'flakes' of N.T.C. oxide semiconductor about 10 μ thick, in the form

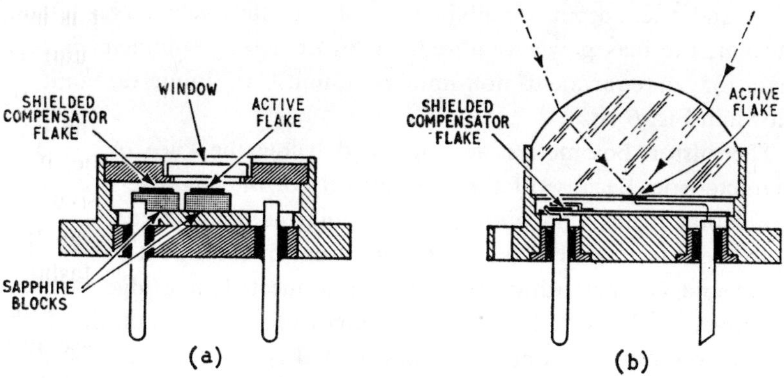

Fig. 6.1. (a) Typical flake bolometer mount for bridge working; (b) flake bolometer immersed in hemispherical lens

of squares or rectangles, with linear dimensions varying from 0·1 to 10 mm, to which end-contacts are made.[7] Resistances usually lie between 1 and 10 MΩ. The upper surface of the exposed element is blackened to achieve maximum emissivity: in practice an absorption of 80% or better can be achieved. The second element is used to compensate the effect of ambient temperature changes on the active, exposed element. The flakes are mounted on insulating substrates such as glass, quartz, sapphire or beryllia, which may be connected directly to a heat sink (solid-backed) as shown, or may be 'gas or vacuum'-backed. For operation as a d.c. detector the latter is the required form since the thermal conductance K is then small and in consequence the responsivity is large. The fact that the thermal time constant $\tau = C/K$ is long is not important for the d.c. mode of operation. For a.c. operation, however, the time constant must be reasonably short to permit operation at useful chopping frequencies. For this mode the solid-backed design is used. The decrease in τ is achieved at the expense of a decrease in $\mathcal{R}$.

Considerations of size and weight place practical limits on the radiation-collecting capability of optical systems. These can be enhanced by immersing the active element (mounting it on the plane surface of a hemispherical lens made from a higher refractive index material such as germanium[8]). This gives optical gain in a small package. The signal/noise ratio is improved, since D^* is that of a smaller flake and the output signal is that of an effectively larger plate. Further, the bias power required for an immersed bolometer is less than for an equivalent non-immersed unit. An immersed unit is shown in Fig. 6.1b.

Thermistor bolometers are operated below the knee of the d.c. characteristic. Because of the non-linearity of the characteristic, it is found that there is an optimum bias point as far as responsivity is concerned[9] (see Section 6.2.4). In the usual dual-element design the active and compensating elements are connected in bridge fashion as shown in Fig. 6.2. Equal bias voltages of opposite polarity with respect to ground are applied across the flakes to minimise noise and microphonics. Because the individual characteristics are accurately matched, both exhibit equal resistance changes when changes in T_a

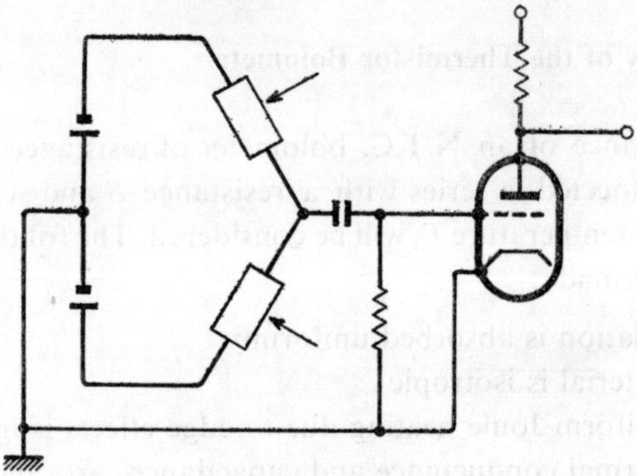

Fig. 6.2. Typical infrared thermistor bolometer detector circuit

occur so that the potential of the common junction remains essentially at ground. With no radiation incident on the detector the bridge is balanced. Absorption of radiation causes an unbalance signal. Typical characteristics of some commercially-available thermistor flake bolometers are given in Table 6.1. For comparison the characteristics of some other detectors are also shown.[3]

Table 6.1. INFRARED BOLOMETER CHARACTERISTICS

Detector	*Operating Temperature* K	*D** cm $Hz^{1/2}$ W^{-1} *Measuring frequency indicated*	*Response Time* τ_c ms	*Optimum Chopping Frequency* Hz
Thermistor	295	$1{\cdot}95\times10^6$ (10 Hz)	1·5	< 30
Thermocouple	295	$1{\cdot}4\times10^9$ (5 Hz)	36	< 5
Golay cell	295	$1{\cdot}67\times10^9$ (10 Hz)	20	< 5
NbN bolometer (superconducting)	15	$4{\cdot}8\times10^9$ (360 Hz)	0·5	

A thermionic valve is used as an unbalance detector if overall signal/noise performance is the main criterion, since the source impedance is high as is the valve input impedance.

6.2.4. Theory of the Thermistor Bolometer

The performance of an N.T.C. bolometer of resistance R and temperature T, connected in series with a resistance S and source of e.m.f. E at ambient temperature T_a will be considered. The following assumptions will be made:

(a) all radiation is absorbed uniformly
(b) the material is isotropic
(c) non-uniform Joule heating due to edge effects is ignored
(d) the thermal conductance and capacitance are independent of frequency.

With no incident radiation the initial heat balance equation is

$$C\frac{\mathrm{d}(T-T_a)}{\mathrm{d}t}+\bar{K}(T-T_a) = I^2R = \frac{E^2R}{(R+S)^2} \tag{6.6}$$

Here $\bar{K}$ represents an average thermal conductance over the range T_a to T. In the steady state

$$\bar{K}(T-T_a) = \frac{E^2R}{(R+S)^2} \tag{6.7}$$

Let radiant power P_r be absorbed, so that T increases by $\Delta T \ll T_0$. The heat transfer equation for the increment becomes

$$C\frac{\mathrm{d}(\Delta T)}{\mathrm{d}t}+K\,\Delta T = \frac{\mathrm{d}}{\mathrm{d}t}(I^2R)\,\Delta T+P_r \tag{6.8}$$

Here K is the thermal conductance at T.

Now
$$\frac{\mathrm{d}}{\mathrm{d}t}(I^2R) = \frac{\alpha E^2R(S-R)}{(S+R)^3} \tag{6.9}$$

Substituting this in Equation 6.8 and also utilising Equation 6.7, we then obtain

$$C\frac{\mathrm{d}(\Delta T)}{\mathrm{d}t}+K_c\,\Delta T = P_r \tag{6.10}$$

where

$$K_c = K-\bar{K}(T-T_a)\alpha(S-R)/(S+R) \tag{6.11}$$

is an effective conductance. It follows that

$$\tau_c = \frac{C}{K_c} = C/\{K - \bar{K}(T - T_a)\alpha(S - R)/(S + R)\} \tag{6.12}$$

is the effective thermal circuit time constant. If P_r is assumed to be periodic at angular frequency ω; i.e.

$$P_r = P_{r0} \exp(\mathrm{j}\omega t) \tag{6.13}$$

then the solution of Equation 6.10 is

$$\Delta T = \Delta T_0 \exp(-t/\tau_c) + \frac{P_{r0} \exp(\mathrm{j}\omega t)}{K_c + \mathrm{j}\omega C} \tag{6.14}$$

The first term on the right-hand side represents a transient and the second a periodic function. The only usable condition is with τ_c positive; i.e. with $K_c(T_0 - T_a)\alpha(S - R)/(S + R) > -K$; otherwise self-burnout occurs (cf. Section 3.2.6). In the steady state, when the transient has decayed

$$\Delta T = \frac{P_{r0}}{(K_c^2 + \omega^2 C^2)^{1/2}} \tag{6.15}$$

6.2.4.1. RESPONSIVITY

If the change in R associated with the change in ΔT is $\Delta R, \ll R$, the corresponding increment of voltage V_s across the load resistance S is

$$V_s = \frac{ES\,\Delta R}{(S + R)^2} = \frac{IS\,\Delta R}{(S + R)} \tag{6.16}$$

where I is the steady current flowing through the thermistor.

For present purposes the responsivity will be defined as the ratio of the open-circuit voltage V_s across S to the amplitude of the radiant power P'_{r0}, falling on the thermistor. Of this the power absorbed is $P_{r0} = \varepsilon P'_{r0}$, where ε is the emissivity of the thermistor. Since $\Delta R = \alpha R\,\Delta T$, it follows from Equation 6.16 that

$$V_s = \frac{\alpha SRI\,\Delta T}{S + R} \tag{6.17}$$

and substitution for ΔT from Equation 6.15 gives

$$V_s = \frac{\alpha SRI\, P_{r0}}{(S+R)\,(K_c^2+\omega^2C^2)^{1/2}} \tag{6.18}$$

Hence

$$\mathcal{R} = \frac{\alpha SRI}{(S+R)\,(K_c^2+\omega^2C^2)^{1/2}} \tag{6.19}$$

When $S \gg R$ the responsivity is maximised with respect to S:

$$\mathcal{R} = \frac{\alpha\varepsilon RI}{(K_c^2+\omega^2C^2)^{1/2}} = \frac{\alpha\varepsilon R^{1/2}\bar{K}(T-T_a)^{1/2}}{(K_c^2+\omega^2C^2)^{1/2}} \tag{6.20}$$

The second form on the right hand side comes from the substitution $I = \{\bar{K}(T-T_a)/R\}^{1/2}$

For d.c. operation, with a bridge circuit,

$$\mathcal{R} = \frac{\alpha\varepsilon RI}{2K_c} \tag{6.21}$$

It can be seen from Equation 6.20 that maximum responsivity occurs when αRI is a maximum. Since R and $|\alpha|$ decrease with increasing I it is clear that there is an optimum operating temperature.

Taking $B = 3\,000$ K and $T_a = 300$ K the optimum value of T is about 315 K. For this value of T, K_c is approximately equal to $1{\cdot}45\bar{K}$. A typical value of R is $10^6\ \Omega$. With $\bar{K} = 10^{-4}$ W/K we have

$$\left.\begin{aligned} K &= 1{\cdot}45\times10^{-4}\ \text{W/K} \\ I &= 0{\cdot}039\ \text{mA} \\ \mathcal{R} &= 8\,000\ \text{V/W} \end{aligned}\right\} \tag{6.22}$$

At high frequencies such that $\omega^2C^2 \gg K_c^2$ Equation 6.20 reduces to

$$\mathcal{R} = \frac{\alpha\varepsilon R^{1/2}\bar{K}(T-T_a)^{1/2}}{\omega C} \tag{6.23}$$

In this range $\mathcal{R}$ is inversely proportional to ω and an increase in $\bar{K}$ now *increases* $\mathcal{R}$, which is the opposite situation to when $\omega^2C^2 \ll K_c^2$. For a given value of ωC there is therefore an optimum value of K_c

which will maximise $\mathcal{R}$. If it is assumed that $K_c \propto \bar{K}$, then the condition for maximum $\mathcal{R}$ is that

$$K_c = \omega C \tag{6.24}$$

If K_c is adjusted for each value of the chopping frequency to satisfy Equation 6.24 then $\mathcal{R} \propto \omega^{-1/2}$. The effect of inhomogenous temperature distribution is considered by Smith et al.[2]

6.2.4.2. DETECTIVITY, D*

This is defined in Equation 6.3 as $A^{1/2}/P_n$, where P_n is the noise equivalent power. It will be assumed that the minimum detectable signal is equal to this. The total mean square voltage $\overline{V_{ne}^2}$ of the electrical noise has been effectively given in Equation 4.61. In the most favourable situation thermal noise alone will be encountered. To this must be added the mean square voltage $\overline{|V_{nt}|^2}$ due to thermal fluctuations. This is given by

$$\overline{|V_{nt}|^2} = \mathcal{R}^2 \overline{|\Delta P_r|^2} \tag{6.25}$$

where $\overline{|\Delta P_r|^2}$ is defined in Equation 4.63. The total mean square noise voltage is

$$\overline{|V_n|^2} = \overline{|V_{ne}|^2} + \mathcal{R}^2 \overline{|\Delta P_r|^2} \tag{6.26}$$

We need now to find the absorbed radiant power which will produce signal voltage V_s equal to $\overline{|V_n|^2}^{1/2}$. Substitution for I as $(\bar{K}(T-T_a)/R)^{1/2}$ in Equation 6.18 gives

$$V_s = \frac{\alpha S R^{1/2} K^{1/2} (T-T_a)^{1/2} P_{r0}}{(S+R)\,(K_c^2+\omega^2 C^2)^{1/2}} \tag{6.27}$$

To determine the detectivity it will be assumed that only Johnson noise is present and that $T \approx T_a$, so that $R \approx R_0$, R_∞ (see Section 4.2.1): then

$$\overline{|V_n|^2} = 4kTR \tag{6.28}$$

for unit bandwidth. Putting $V_s = 2(kTR)^{1/2}$ in Equation 6.27 we find that

$$P_n = \text{N.E.P.} = \frac{(S+R)\,(K_c^2+\omega^2 C^2)^{1/2}\, 2(kTR)^{1/2}}{\alpha S R^{1/2} \bar{K}^{1/2} (T-T_a)^{1/2}} \tag{6.29}$$

From Equation 6.3 it follows that

$$D^* = \frac{A^{1/2}\alpha S R^{1/2}\bar{K}^{1/2}(T-T_a)^{1/2}}{2(kTR)^{1/2}(S+R)(K_c^2+\omega^2C^2)^{1/2}} \tag{6.30}$$

Making the further assumptions that $\omega C \ll K_c$, $S \gg R$, the open-circuit detectivity becomes

$$D^* = \frac{A^{1/2}B(T-T_a)^{1/2}}{2k^{1/2}T^{5/2}\bar{K}^{1/2}(1+(T-T_a)B/T^2)} \tag{6.31}$$

where substitution has been made for K_c from Equation 6.11, with the assumption that $K \approx \bar{K}^{1/2}$ and $\alpha = -B/T_a^2$. $\bar{K}$ can be written as A/k_0d, where k_0 is the thermal conductivity of the layer between the thermistor flake and the heat sink and d is its thickness. Hence D^* can be written as

$$D^* = \frac{(T-T_a)^{1/2}Bd^{1/2}}{2k^{1/2}T^{5/2}k_0^{1/2}[1+(T-T_a)B/T^2]} \tag{6.32}$$

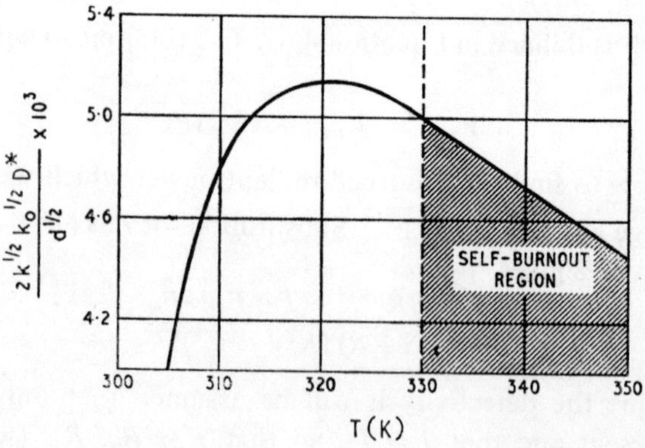

Fig. 6.3. Dependence of constant × D on operating temperature (after Kruse et al)*

It is to be noted that D^* is independent of wavelength, detector area and resistance. The relative dependence of D^* on T for $T_a = 300$ K and $B = 3\,600$ K is shown in Fig. 6.3. Self-burnout occurs for $T \geq 330$ K.

6.3. MICROWAVE POWER MEASUREMENT

6.3.1. Introduction

At power frequencies instruments exist by means of which current, voltage and power may all be measured; i.e. a quantitative evaluation may be made in terms of standard electrical quantities. At lower radio frequencies similar measurements may be made, although there is increasing error in current and voltage measurements as the dimensions of the measuring instruments and interconnections become significant fractions of the wavelength* of the electromagnetic field being investigated. Beyond about 1 GHz current and voltage measurements are not made; power and impedance are of fundamental significance. At relatively high power levels, in excess of 1 W, the water calorimeter has generally been regarded as a standard instrument for the measurement of microwave power[10] and this has been augmented by the torque-operated wattmeter, which is dependent on radiation pressure. Between 10 mW and 1 W, calibrated attenuators and directional couplers may be used to bring the power down to the low level range of 1 μW–10 mW. In this range metal-wire and thermistor bolometers are used for detection and measurement at frequencies up to some 30 GHz, although there are significant errors in measurements at the higher frequencies.[11–15] The direct measurement of power by substitution relies on the equivalence of the r.f. power being investigated and a known amount of d.c. or l.f. power in producing a significant temperature rise in the sensing element. This in turn necessitates the equivalence of the resistance of the element at the calibration and measuring frequencies. Complete equivalence does not arise because the spatial distributions of current, power and resistance within the thermistor are different for the r.f. and direct powers. Even though d.c. and microwave resitances are not equivalent, however, calibration

* The approximate relationship between wavelength λ and frequency f is

$$\lambda = \frac{3\times10^4}{f}\ \text{cm}$$

where f is in MHz; i.e. at a frequency of 3 GHz $\lambda = 10$ cm may be made against a standard instrument, so that the frequency range in which thermistor bolometers can be used is extensive.

Errors arise because the temperature distribution is not identical in the calibration and measurement conditions and a fraction of the microwave power is dissipated in parts of the mount housing the sensing element, such as the dielectric coating around the element itself, the supporting wires and other metallic surfaces, particularly those in any associated impedance transformers.

All thermistor mounts share the desirable 'self-fusing' property. The burnout resistance of a thermistor ($\approx 5\ \Omega$) is so much lower than the operating resistance (≈ 100–$200\ \Omega$) that severe mismatch occurs before the thermistor can be burned out. This is not so for metal-wire bolometers.

6.3.2. Frequency Dependence of Thermistor Resistance

The type of thermistor in general use for microwave power measurements is the type *E* described in Section 1.2. The lengths of the thin internal lead wires mean that their inductances cannot be neglected at microwave frequencies. Their resistance is dependent on skin-effect, and will be proportional to $\sqrt{f}$ in this frequency range. The equivalent circuit of the bead itself has not been fully evaluated. It has been suggested that it may be represented by a three-element equivalent circuit comprising a resistive path of resistance R_1, representing terminal resistance between granules in intimate contact, in parallel with a series R_2 and C_2 combination representing the terminal impedance between granules which are capacitively coupled. On the other hand Collard et al[11] have assumed a simple parallel *CR* combination to represent the bead. Since their results appear to be the only quantitative ones available in the open literature* their model will be adopt-

* Walker[16] has considered the bead alone to be represented by a uniformly-distributed *R–C* network in the u.h.f. range and has shown that measured values of u.h.f. resistance (at about 400 MHz) agree resonably well with predicted u.h.f. values, which are more than an order of magnitude smaller than the d.c. resistance. The thermistors used had d.c. resistances of 73 kΩ and 8·9 MΩ: for such high-resistance thermistors capacitive effects are much more marked than for the $\approx 100\ \Omega$ thermistors used for microwave power measurement.

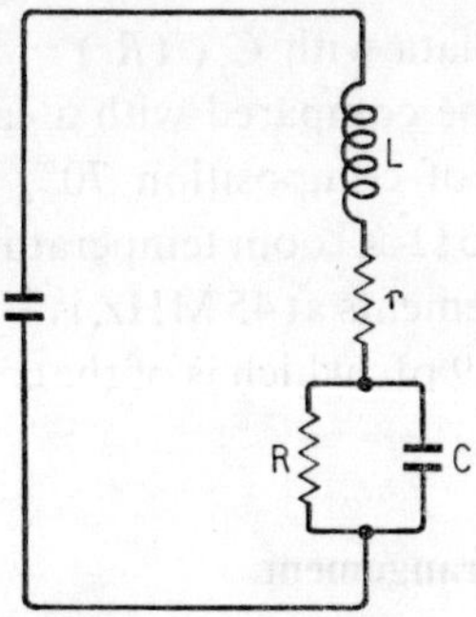

Fig. 6.4. Equivalent circuit of a type-E thermistor

ed here. The full equivalent circuit incorporating this is shown in Fig. 6.4. L represents the lead inductance, r the lead resistance and C_s the external shunt capacitance due to the encapsulating glass bead and mount. The latter can be absorbed in the external circuit. From simple circuit considerations it can be shown that the ratio, β, of the power dissipated in the bead to the total power absorbed is given by

$$\beta = \frac{1}{1+\omega^2C^2Rr} \tag{6.33}$$

if $R \gg r$. If r is assumed to be proportional to $\sqrt{f}$ and R and C independent of frequency, then

$$\beta = \frac{1}{(1+Af^{2\cdot5})} \tag{6.34}$$

where A is a constant for each bead. In Fig. 6.5 Equation 6.34 is

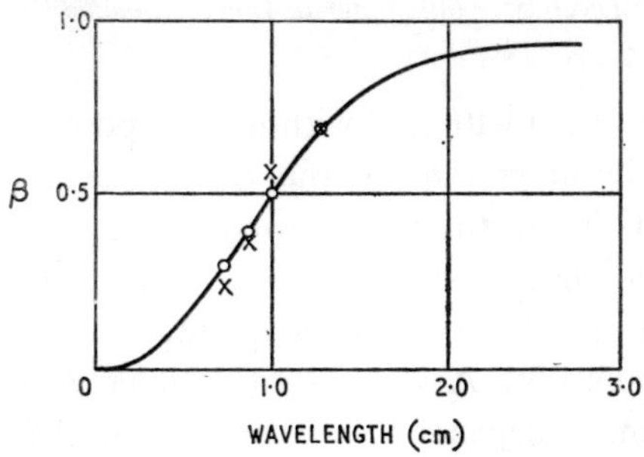

Fig. 6.5. Fit of experimental data relating β to wavelength using Equations 6.33 and 6.34: o against enthrakometer, x against calorimeter. (after Collard et al)

fitted to experimental data with $C\sqrt{(Rr)} = 5{\cdot}3$ pF at $\lambda = 1$ cm. This value of $C\sqrt{(Rr)}$ can be compared with a calculated value. For 0·001 in diameter lead wire of composition 70% Pt and 30% Ir, 0·04 in long, the value of r is 2·5 Ω at room temperature. An approximate value for C, based on measurements at 45 MHz, is ~0·2 pF so with $R = 150\,\Omega$ this gives $C\sqrt{(Rr)} = 3{\cdot}9$ pF, which is of the correct order of magnitude.

6.3.3. Basic Circuit Arrangement

The thermistor is usually connected in a Wheatstone bridge circuit[12] as shown in Fig. 6.6. The bridge may be operated either in a balanced or unbalanced (direct-reading) condition. In the former case the

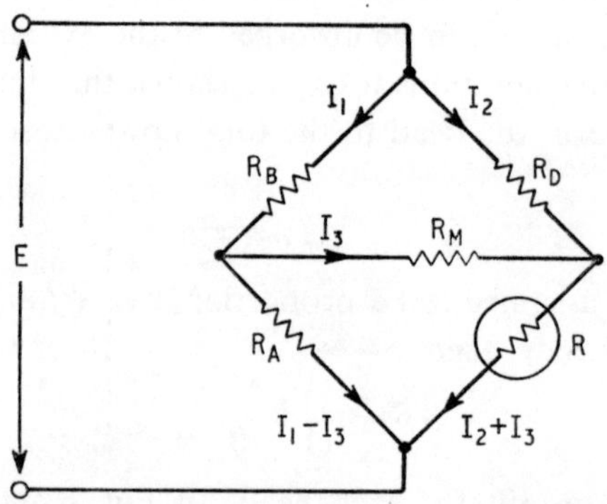

Fig. 6.6. Thermistor bolometer in Wheatstone bridge circuit

bridge is balanced both with and without r.f. power dissipation in the thermistor and the magnitude of the r.f. power is determined from the difference in thermistor bias power in the two conditions. The out-of-balance voltage of an unbalanced bridge is used to indicate the absorbed r.f. power. Unbalanced bridges are usually calibrated against a balanced bridge. A well-designed direct-reading bridge has an approximately linear dependence of output on absorbed r.f. power. The fundamental principles of bridge design are common to both balanced and unbalanced bridges.

It is necessary to distinguish between thermistor sensitivity and bridge sensitivity. Thermistor sensitivity is defined by Equation 3.5 or sometimes simply as dR/dP: in the latter case it is usually expressed in $\Omega\ mW^{-1}$. Bridge sensitivity is a measure of the bridge unbalance resulting from the absorption of r.f. power by the thermistor. It is often expressed as $\mu A\ mW^{-1}$. It is not necessarily true that the thermistor with the highest sensitivity will provide the greatest bridge sensitivity in a given bridge circuit.

In Fig. 6.6 the thermistor resistance is R; R_A, R_B and R_D are the resistances of the other arms, assumed temperature insensitive and R_M, also assumed temperature insensitive, is the resistance of the meter used to detect the balance or out-of-balance condition. The relevant circuit equations are

$$I_1R_B+I_3R_M = I_2R_D \tag{6.35}$$

$$(I_1-I_3)R_A = I_3R_M+(I_2+I_3)R \tag{6.36}$$

$$I_1R_B+(I_1-I_3)R_A = E \tag{6.37}$$

If it is assumed that $R_A = R_D = R_B$ and that $R = R_B-\Delta R$, with $\Delta R \ll R_B$, then the following expressions are obtained for the direct currents flowing;

$$I_3 = \frac{E\Delta R}{4R_B(R_B+R_M)}\left[1+\frac{\Delta R}{4(R_B+R_M)}\,\frac{(3R_B+R_M)}{R_B}\right] \tag{6.38}$$

$$I_2 = \frac{E}{2R_B}\left[1+\frac{\Delta R}{4(R_B+R_M)}\,\frac{(R_B+2R_M)}{R_B}\right] \tag{6.39}$$

$$I_1 = \frac{E}{2R_B}\left[1+\frac{\Delta R}{4(R_B+R_M)}\right] \tag{6.40}$$

$$I_2+I_3 = \frac{E}{2R_B}+\frac{\Delta R E(3R_B+2R_M)}{8R_B^2(R_B+R_M)} \tag{6.41}$$

These equations apply for a constant driving voltage at the bridge terminals (voltage source). As $\Delta R \ll R$, it is permissible to write

$$\Delta R = \frac{dR}{dP}\,\Delta P \tag{6.42}$$

where ΔP is the absorbed r.f. power and dR/dP is the thermistor sensitivity. From Equations 6.38 and 6.42 the bridge sensitivity may be written as

$$\frac{I_3}{\Delta P}=\frac{E\,\mathrm{d}R/\mathrm{d}P}{4R_B(R_B+R_M)}\left[1+\frac{(\mathrm{d}R/\mathrm{d}P)\,\Delta P(3R_B+R_M)}{4R_B(R_B+R_M)}\right] \qquad (6.43)$$

For small meter deflections in sensitive bridge circuits, the second erm in the square brackets may be ignored, so giving

$$\frac{I_3}{\Delta P}=\frac{E\,\mathrm{d}R/\mathrm{d}P}{4R_B(R_B+R_M)} \qquad (6.44)$$

From equation 3.6 it follows that

$$\left|\frac{\mathrm{d}R}{\mathrm{d}P}\right|=\frac{BR}{K(P/K+T_a)^2} \qquad (6.45)$$

so that the expression for bridge sensitivity becomes

$$\frac{I_3}{\Delta P}=\frac{EBR}{4R_B(R_B+R_M)\,K(P/K+T_a)^2} \qquad (6.46)$$

At balance $R=R_B$ so that

$$\frac{I_3}{\Delta P}=\frac{EB}{4(R+R_M)\,K(P/K+T_a)^2} \qquad (6.47)$$

For good bridge sensitivity it is clear that a meter should be chosen with large current sensitivity and low resistance. E cannot be made too large because of the limitation imposed by the maximum power dissipation of the thermistor. In some instances additional sensitivity may be gained if the bridge arms are not equal, but the gain is seldom large. Bridge circuits with unequal arms are usually designed to meet other requirements, such as reduced bridge current drain or greater meter protection in the event of thermistor burnout.

In the discussion above it has been assumed that the bridge is fed from a voltage source. If the supply resistance is not negligible then E will decrease somewhat with increasing ΔP, but the effect is negligible for small unbalance of the bridge.

Bridge sensitivity can be increased by shunting the thermistor with a temperature-insensitive resistor. The smaller the shunting resistance, the greater the increase in bridge sensitivity. Although there is no theoretical limit to the sensitivity that can be achieved in this way it is usual to design for modest gains only. The reasons are that with heavy shunting, (a) the effective thermal time constant of the circuit becomes long so that the bridge response is sluggish (see Section 4.2.3); (b) there is a marked departure from linearity and (c) the total bridge current needed is greatly increased. The explanation of the increased sensitivity achieved by shunting is that it lowers the effective source resistance presented to the thermistor. Hence, when microwave power is absorbed, so reducing R, there is also an increase in the direct power absorbed, producing an augmented temperature rise and further fall in R (positive feedback).

6.3.3.1. LINEARITY OF THE DIRECT-READING BRIDGE

Ideally it is desirable that the out-of-balance indication be linearly related to the r.f. power absorbed. It follows from Equation 6.44 that the prime requirement is that $\mathrm{d}R/\mathrm{d}P$ is independent of P. It has been demonstrated that this is not so for thermistors. In practice, however, it is found[17] that for a direct-reading bridge, giving full-scale indication for less than 200 μW of incident r.f. power, the departure from linearity is less than $\pm 2\%$ and for powers up to 1 mW is less than $\pm 5\%$. Other factors which affect the linearity of the bridge are

(a) the second term in the square brackets of Equation 6.43
(b) impedance mismatch at microwave frequencies, due to change in R from the initially matched value
(c) non-linearity of the output current meter
(d) shunting the thermistor (see previous section)

6.3.3.2. TEMPERATURE-COMPENSATION OF THE DIRECT-READING BRIDGE[12]

In a practical direct-reading bridge circuit it is desirable that the bridge sensitivity should not be dependent on changes in ambient temperature and also that departure from balance (drift) should not occur: drift is unbalance caused by change in ambient temperature. Temperature compensation of bridge sensitivity is of prime importance, compensation for drift being for convenience only. It should be noted, however, that whereas the problem of compensation for sensitivity change is virtually independent of bridge sensitivity, the degree of drift increases directly with bridge sensitivity.

Experimentally it is found that the sensitivity of an uncompensated thermistor bridge, in which balance is brought about by altering the bridge current, increases as the ambient temperature decreases. It is also found experimentally that bridge sensitivity varies directly with the bridge terminal voltage, as this is changed by adjusting the bridge current. Hence, to stabilise bridge sensitivity as the ambient temperature changes a near-constant bridge voltage should be maintained.

One of the first compensated systems made is known as the *V* bridge. This was designed to operate a bead thermistor at a resistance between 200 and 300 Ω. Cascaded voltage regulator tubes were used to provide a well-stabilised d.c. supply, which produced sufficient direct current to balance the bridge at high temperatures, but insufficient at low temperatures. This supply is therefore augmented by an audio frequency source to achieve balance at low temperatures. A d.c. meter is used as the bridge galvanometer: since this does not respond to the a.f. signal and the direct bridge voltage is kept virtually constant, the bridge sensitivity is relatively independent of temperature. The compensation gives ±0·25 dB variation of sensitivity over a temperature range from −10 °C to +60 °C, compared with an uncompensated sensitivity variation in excess of ±1 dB. Drift compensation in this design is achieved with an N.T.C. disc thermistor connected in parallel with the bridge in the a.f. supply. As the ambient temperature increases the resistance of the compensating thermistor decreases, thereby increasing its shunting effect on the a.f. supply, so decreasing the a.f. power applied to the bead thermistor and tending to keep its resistance constant.

Greater sensitivity than is possible with the essentially d.c.-type *V* bridge is achieved in the so-called *W* and *X* bridges in which the initial bridge balance is obtained entirely with an a.f. power source. In operation the bridge a.c. unbalance signal is amplified (the source of increased sensitivity), phase-detected and registered by a d.c. milliammeter. Phase-detection is necessary to indicate whether the resistance of the thermistor bead is greater or less than that required for bridge balance. A d.c. calibrating signal is used to standardise the bridge sensitivity at the working ambient temperature; this obviates the need for a sensitivity compensator. Drift-compensating networks at high sensitivities are not effective because of the slight temperature differences that inevitably exist between the detector thermistor and the compensating network. Control of drift is achieved by enclosing the thermistor in a thermostatically-controlled oven or other suitable constant-temperature enclosure. Full scale deflection of the *W* bridge is obtained for 0·1 mW of microwave power and for the *X* bridge, which is more complicated, for 15 μW. Full details of *V*, *W* and *X* bridges are given by Griesheimer.

6.3.3.3. SELF-BALANCING BRIDGES[12, 18–20]

Balancing and rebalancing procedures and the associated calculations are time-consuming. Because of this, self-balancing bridges have been developed in which the r.f. power absorbed is automatically compensated by a decrease in a.f. bias power, which is metered and therefore registers the r.f. power. A schematic example of a self-balancing bridge is given in Fig. 6.7. When the system is first turned on, with the thermistor at ambient temperature, the bridge is unbalanced. There is positive-feedback between the output and input of the a.f. amplifier which causes the amplifier to oscillate and supply a.f. power to the bridge. The amplitude of the oscillation decreases as the

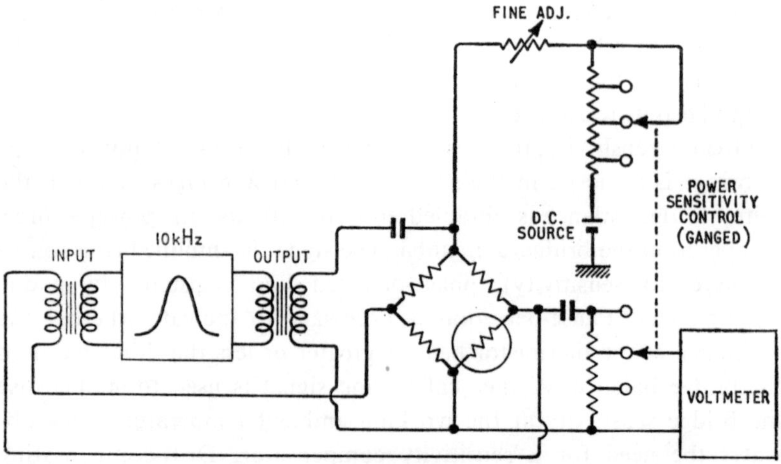

Fig. 6.7. Schematic circuit of a self-balancing thermistor bridge

thermistor is warmed up, until finally the loop gain of the feedback network becomes unity. For an amplifier gain of 1 000 the thermistor resistance is then within 0·1% of the balance value. A high impedance voltmeter, responding to a.f., connected across the thermistor, meters the a.f. power supplied at the initial balance and is adjusted to give zero indication in this condition. When microwave power is absorbed, the thermistor is heated and its resistance falls, unbalancing the bridge. The unbalance is amplified and fed back to the bridge to restore balance, the a.f. power reading giving, by substitution, the value of the absorbed microwave power.

Aslan's[19] design incorporated two bridges in each of which an identical thermistor comprised one arm: one bridge was used for r.f. sensing and the other to provide temperature compensation.

6.3.3.4. TWO-DISC THERMISTOR BRIDGE[12]

A more simple bridge requiring only a d.c. supply is the two-disc thermistor bridge. In this circuit sensitivity-compensation is achieved by incorporating a disc thermistor, embedded in a resistive network, in series with the bridge meter. Here use is made of the fact that bridge sensitivity depends on the resistance R_M of the meter arm (see Equation 6.47) and the ability to shape the resistance-temperature curve of a thermistor using temperature-insensitive passive resistors (see Section 3.2.8). The circuit diagram of this bridge is shown in Fig. 6.8. Component values for different bridge designs are given in Table 6.2.

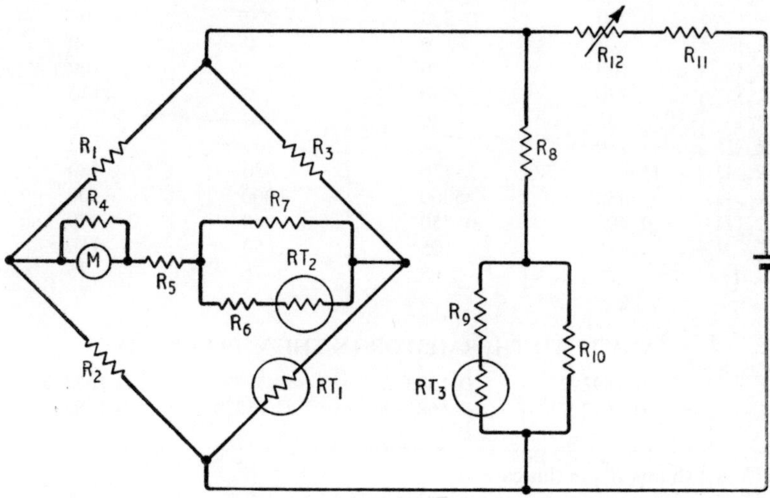

Fig. 6.8. Detailed circuit diagram of the two-disc thermistor bridge

6.3.4. Dual Element Circuits[20]

In the design of coaxial thermistor mounts using the theoretical basis so far discussed, with a single thermistor incorporated in a Wheatstone bridge circuit, there is a practical complication, since d.c. connections must be made to both the inner and outer conductors, between which the thermistor is connected. This is accomplished by using a stub to provide both an a.c. and d.c. short circuit at the point of connection, but an a.c. open circuit at the connection of the stub to the main line.

Another way of making a coaxial thermistor bolometer, is to connect two thermistors between inner and outer conductors in a

Table 6.2. CIRCUIT CONSTANTS FOR FIG. 6.7**

Component	*Circuit 1*	*Circuit 2*	*Circuit 3*‡	*Circuit 4*
$R_1\,\Omega$	430	250	300	180
$R_2\,\Omega$	430	250	300	180
$R_3\,\Omega$	180	250	130	130
$R_4\,\Omega$	0–200	0–500	0–200†	0–200†
$R_5\,\Omega$	82	0	0	0
$R_6\,\Omega$	120	20	0	0
$R_7\,\Omega$	270	250	0	130
$R_8\,\Omega$	33	40	0	0
$R_9\,\Omega$	0	36	200	82
$R_{10}\,\Omega$	15 000	475	620	780
$R_{11}\,\Omega$	330*	5000	7 000	6 500
$R_{12}\,\Omega$	0–400	0–750	0–1 000	0–1 000
E V	6	105	150	150
$RM\,\Omega$	75	68	75	75
T_1	ANY BEAD THERMISTOR IN THE V519 FAMILIY §			
T_2	*D*-169227	*D*-168392	—	*D*-168392
T_3	*D*-169227	*D*-168391	*D*-168392	*D*-168391

* R_{11} shunts R_{12} in this case
† R_4 is in series with the meter in this case
‡ example of a one-disc thermistor bridge
§ Bell Telephone Laboratories
** After Montgomery, 'Technique of Microwave Measurements', McGraw Hill (1947) (Radiation Laboratory Series 11, p. 118)

terminating cavity by a *T*-section. The thermistors are in series as far as the d.c. or a.f. power supply is concerned, being isolated from the inner and outer connectors by blocking and lead-through capacitors respectively: to the radiant power they are effectively in parallel and their parallel resistance is therefore chosen to equal the characteristic impedance of the measuring line.

This technique gives rise to an additional source of error[21] since the two thermistors may not have equal resistances. If this is so the thermistor with the higher resistance will dissipate more direct power than the other, while the thermistor with the lower resistance will

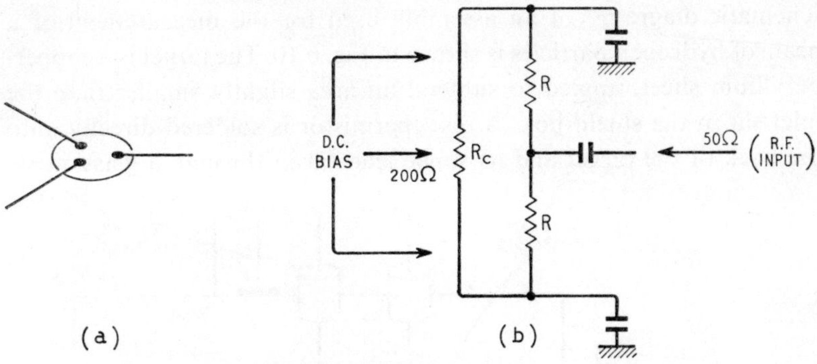

Fig. 6.9. (a) Centre-tapped thermistor, (b) equivalent electrical circuit

dissipate more r.f. power. In consequence the substituted direct power will not equal the total absorbed r.f. power, unless the dissipation constants of both thermistors are identical. The use of a single centre-tapped thermistor bead has been suggested to obviate this source of error.[22] Since only one very small bead is used it is reasonable to assume that a common dissipation constant is valid. The form of the tapped bead thermistor is shown in Fig. 6.9a and its lumped equivalent circuit in Fig. 6.9b. The latter incorporates a parallel resistive component R_c, in view of the geometry.

6.3.5. Practical Thermistor Mounts

The dual-element thermistor mount is commonly used in coaxial instruments. Practical designs are given in the literature.[10, 12, 19, 22] Waveguide mounts are usually designed for single thermistors and incorporate impedance transformers to bring the thermistor impedance to the waveguide impedance.

6.4. PARTICLE BEAM MEASUREMENT

A calorimetric method of measuring the energy of a particle beam has the advantage that it is charge independent so that it can be used for beams composed of either neutral or ionised particles. The schematic diagram[23] of an assembly used for the measurement of a beam of hydrogen particles is shown in Fig. 6.10. The target is a copper-beryllium sheet, angled to subtend an area slightly smaller than the inlet slit in the shield-box. A disc thermistor is soldered directly onto the back of the target and its single lead taken through a glass–metal

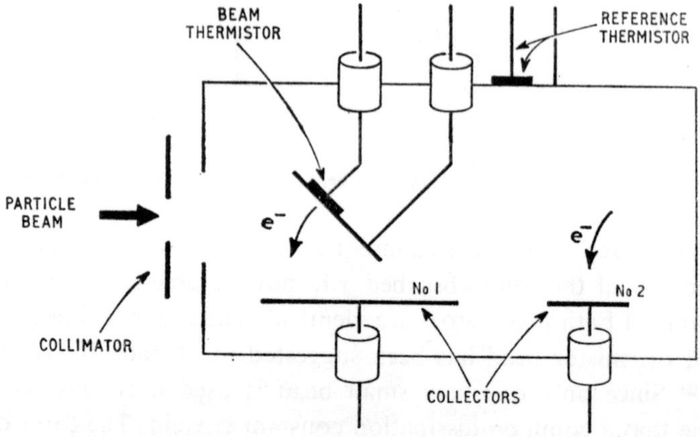

Fig. 6.10. Schematic diagram of apparatus used for measuring the energy of hydrogen particles

seal to a bridge outside the box: the second bridge lead is a fine wire making ohmic contact to the target. The shield box, made of 1/8 in thick copper, has a relatively large heat capacity and is relatively insensitive to ambient temperature fluctuation. Its temperature is monitored by a second thermistor soldered to its outside surface. Secondary electron emission from the target is measured by collector 1 (+60 V bias) and secondary emission from the wall, caused by incident particles missing the target, by collector 2. The thermal time constant of the target is 30 s.

For such a calorimeter the heat balance equation is the same as that for a thermistor alone; i.e.

$$\frac{dQ}{dt} = C\frac{d(\Delta T)}{dt} + G\,\Delta T \tag{6.48}$$

Here Q is the heat equivalent of the particle beam, C the thermal capacity of the target and G its thermal conductance. The general solution of this equation for ΔT is

$$\Delta T = \frac{1}{G}\frac{dQ}{dt} + \left[\Delta T(O) - \frac{1}{G}\frac{dQ}{dt}\right] \exp(-t/\tau) \qquad (6.49)$$

where $\tau = C/G$ is the thermal time constant of the target and $\Delta T(0)$ is the value of ΔT at $t = 0$. The initial rate of temperature rise is found to be

$$\frac{d(\Delta T)}{dt} = \frac{1}{C}\frac{dQ}{dt} \qquad (6.50)$$

for $\Delta T(0) = 0$; i.e. the initial rate of change of target temperature is proportional to the incident power. Hence if the thermal capacity of the target is known the incident power can be calculated from the initial rate of temperature rise. Alternatively the incident power may be obtained from the steady-state temperature rise (first term on the right hand side of Equation 6.49) if G is known.

REFERENCES

1. MELMAN, I. J. and MELTZER, I. M. 'Status report on infrared thermistor detectors', *Proc. Nat. Electron. Conf.*, **18,** 556 (1962)
2. SMITH, R. A., JONES, F. E., and CHASMAR, R. P. *The detection and measurement of infrared radiation*, Oxford University Press (1957)
3. KRUSE, P. W., MCGLAUCHLIN, L. D., and MCQUISTAN, R. B. *Elements of infrared technology: generation, transmission and detection*, Wiley (1962)
4. FLORANT, L. E. 'Testing using infrared' ,*J. Instrum. Soc. Am.*, **11,** 61 (1964)
5. BARNES, R. R. and LANE, W. Z. 'Thermography: diagnostic infrared instrumentation', *Proc. a. Conf. Engng Med. Biol.*, **5,** 50 (1963)
6. JONES, R. C. 'Factors of merit for radiation detectors', *J. Opt. Soc. Am.*, **39,** 344 (1949)
7. WORMSER, E. M. 'Properties of thermistor infrared detectors', *J. Opt. Soc. Am.*, **43,** 15 (1953)
8. BARNES, R. R. 'Thermistor infrared detectors', *Bulletin 200, Barnes Engineering Company Defense and Space Division Stamford, Connecticut.*
9. JONES, R. C. 'Steady State Load Curves for Semi-Conductor Bolometers', *J. Opt. Soc. Am.*, **36,** 448 (1946)
10. GINZTON, E. L. *Microwave measurements*, McGraw-Hill (1957) Chapter 3
11. COLLARD, J., NICOLL, G. R., and LINES, A. W. 'Discrepancies in the measurement of microwave power at wavelengths below 3cm, *Proc. Phys. Soc.*, **63,** 215 (1950)

12. MONTGOMERY, C. G. *Technique of microwave measurements (M.I.T. Radiation Laboratory Series)*, McGraw-Hill (1947)
13. LANE, J. A. 'The measurement of power at a wavelength of 3cm. by thermistors and bolometers', *Proc. Instn. elect. Engs.*, **102B**, 819 (1955)
14. STREET, R. 'The absolute measurement of low power at 3000 MHz, *Proc. Instn. elect. Engrs.*, **96A–2**, 391 (1949)
15. STREET, R. and WHITAKER, P. D. 'Measurement of microwave power at wavelengths of 3cm and 10cm', *Proc. Phys. Soc.*, **63**, 623 (1950)
16. WALKER, J. 'Thermistors at high frequencies' *Wireless Engr*, **24**, 28 (1947)
17. BATH, C. C. and GOLDBERG, H. 'Self-balancing thermistor bridge', *Proc. Natn Electron. Conf.*, **3**, 47 (1947)
18. HANSEN, W. W., WOODYARD, J. R., and GINZTON, E. L. 'Automatic power bridge' *U.S. Patent No. 2,525,901*
19. ASLAN, E. E. 'Temperature-compensated microwatt power meter' *I.R.E. Trans. Instrumn.*, **9**, 291 (1960)
20. GAINSBOROUGH, G. F. *Technique of microwave measurements (M.I.T. Radiation Laboratory Series)*, 136, McGraw-Hill (1947)
21. ENGEN, G. F. 'A d.c.-r.f. substitution error in dual-element bolometer mounts', *I.R.E. Trans. Instrumn*, **13**, 58 (1964)
22. ASLAN, E. E. 'A single-bead broadband coaxial thermistor mount', *I.R.E. Trans. Instrumn*, **14**, 102 (1965)
23. CHAMBERS, E. S. 'Power calorimetry of fast-particle beams' *Rev. Scient. Instrum.*, **35**, 95 (1964)

CHAPTER 7

Applications Based on Variation of Dissipation Constant

7.1. INTRODUCTION

The dissipation constant or thermal conductance, K, is not a property of the thermistor alone, although it is affected by its geometry. It is markedly influenced by the ambient atmosphere. As is well-known, heat can be lost from a body by conduction, convection and radiation. As far as thermistors are concerned the 'atmosphere', for applications involving variation of dissipation constant, is either liquid or gaseous. These applications are diverse and cover many areas of scientific interest and engineering practice. They may be summarised as in the following subdivisions[1]; anemometers, flow-meters and fluid-velocity meters; manometers (vacuum gauges); thermal-conductivity analysis, gas-chromatography; liquid-level measurement and control.

It is obvious that to obtain an indication of environmental change the thermistor should be operated in its non-linear region, such that its operating temperature exceeds that of the ambient atmosphere.

In the following sections the underlying physical principles involved in the applications will be outlined; then generalised examples of basic instrument design will be presented. In conclusion three specific designs will be considered in detail.

Because aged N.T.C. thermistors retain their calibration better than P.T.C. thermistors, only the former will be considered in this chapter.

7.2. EFFECT OF AMBIENT ON *K*

At the atomic or molecular level, heat exchange between a solid body and a surrounding gaseous or liquid ambient atmosphere is the result of collisions of the free particles of the surrounding medium with the solid. The rate at which heat is lost from such a body at a higher temperature than its surrounding medium clearly depends on whether this is stationary or moving. In the latter case the collision rate is higher and the rate of heat loss is correspondingly increased. For gaseous atmospheres the thermal conductivity of the gas increases linearly with pressure (from near vacuum conditions) up to an air-equivalent pressure of about 200 μ, in the so-called molecular-flow range, but becomes independent of pressure at pressures somewhat in excess of this when the probability of molecular collisions increases significantly.

7.2.1. Applications Based on the Rate of Flow

These include anemometers and flow and fluid-velocity meters. Rasmussen's[2] approximate analysis will be followed here to indicate the interaction between *K* and the medium. His interest was in producing an improved bathythermograph and an accurate oceanographic velocity meter, so both ambient temperature T_a and fluid velocity are considered as variables. The starting point is the general heat transfer equation, repeated here;

$$C\frac{dT}{dt}+K(T-T_a) = P \tag{7.1}$$

The approximations made are;

(a) the initial bead temperature is T_0, the corresponding thermistor resistance is R_0 and

$$R = R_0[1+\alpha_0(T-T_0)] \tag{7.2}$$

α_0 being the temperature coefficient of resistance at T_0, (valid if $T - T_0 \ll T_0$);

(b) there is a constant heating current I.

On substitution in Equation 7.1 for $P = I^2R$, with R defined as in Equation 7.2, the following equation is obtained

$$\frac{dT}{dt} + \frac{T}{\tau'} = m + \frac{T_a}{\tau} \qquad (7.3)$$

where

$$\frac{1}{\tau'} = \frac{(K - P_0\alpha_0)}{C}$$

$$m = \frac{P(1 - \alpha_0 T_0)}{C}$$

$$\frac{1}{\tau} = \frac{K}{C}$$

with K a function of velocity, v, and τ', τ and T_a functions of time.

The usefulness of Equation 7.3 may be demonstrated by determining the behaviour in a simple case in which the ambient temperature gradient $g = dT_a/ds$ is constant. It is assumed that a steady state exists at $t = 0$, so that $P_0 = K(T_0 - T_a)$ is the initial condition. The solution of Equation 7.3 is

$$T = T_0 + \frac{\tau'^2}{\tau} gv(e^{-t/\tau'} - 1) + \frac{\tau'}{\tau} gvt \qquad (7.4)$$

It is important to note that the reduction in effective time constant represented by τ'/τ is the result of the dissipation of power in the thermistor, and is specified by

$$\frac{\tau'}{\tau} = \frac{1}{\{1 + BP_0/T_0^2K(v)\}} \qquad (7.5)$$

The reduced time constant, which means a more rapid response, is accompanied by a reduced sensitivity to fluid temperature variations and an increased sensitivity to speed variations.

A knowledge of the dependence of K on the relative speed, v, of the fluid is necessary for a more general solution of Equation 7.3.

Under steady-state conditions we have (Equation 1.1)

$$T = T_a + \frac{P}{K(v)} \tag{7.6}$$

T can also be expressed as follows, from the basic thermistor resistance-temperature Equations (2.9 and 2.11)

$$T = \frac{\alpha_a T_a^2}{\alpha_a T_a - \log_e \left(\frac{R}{R_a}\right)} \tag{7.7}$$

These two expressions can be equated and adjusted to give

$$K(v) = \frac{P}{T_a}\left\{\frac{\alpha_a T_a}{\log_e \left(\frac{R}{R_a}\right)} - 1\right\} \tag{7.8}$$

Hence $K(v)$ can be determined from easily-measured variables. A suitable experimental arrangement for determining $K(v)$ from this equation is outlined by Rasmussen[2]. The thermistor is supported at the tip of a horizontal arm from a vertical shaft, which can be rotated at known velocities within a large Dewar jar containing the fluid in question.

In Fig. 7.1 voltage-current characteristics are shown for a type 51A1 thermistor for various water speeds. Corresponding data are

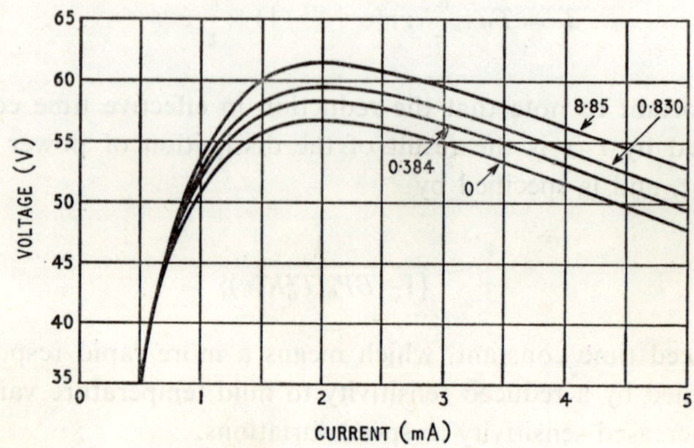

Fig. 7.1. Voltage-current characteristics for a type 51A1 thermistor; water speed is the parameter (after Rasmussen)

given by Rasmussen for air: in the case of air the zero speed value of $K(v)$, $K(0)$, is approximately 1/5th of that in water. In water K rises very rapidly with increasing speed to an asymptotic value about 18% higher than $K(0)$. A good approximation to the practical dependence of K on velocity is given by

$$K(v) = K(0) + a(1 - e^{-bv}) \tag{7.9}$$

where $K(0)+a$ is the limiting value achieved at high speeds and $b = (1/a)\,(dK/dv)v = 0$. The accuracy of this relationship may be seen in Fig. 7.2.

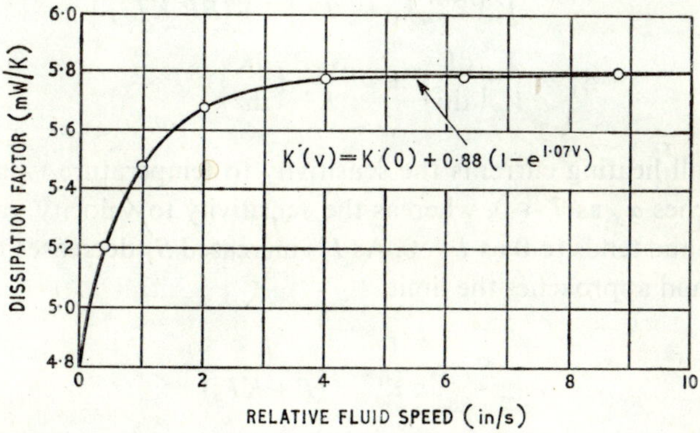

Fig. 7.2. Dependence of K on v for a type 51A1 thermistor in water; the continuous curve is specified by $K(v) = K_0 + 0{\cdot}88\,\{1 - \exp(-1{\cdot}07v)\}$; *circles represent experimental points (after Rasmussen)*

7.2.1.1. TEMPERATURE AND SPEED SENSITIVITIES AT CONSTANT CURRENT

A knowledge of these is important for instrument design. Each is defined as the fractional rate of change of thermistor voltage with respect to one of the variables, with the remaining variable held constant. They may be readily determined from the steady-state heat transfer equation, rewritten here in the form

$$I^2 R_a \exp\left\{B\left(\frac{1}{T} - \frac{1}{T_a}\right)\right\} - K(v)(T - T_a) = 0 \tag{7.10}$$

and considering T as a function of the parameter I and the variables T_a and v, as follows;

$$\frac{\partial T}{\partial T_a} = \frac{1}{1+(BP/KT^2)} \tag{7.11a}$$

$$\frac{\partial T}{\partial v} = -(\partial T/\partial T_a)(P/K^2)(\mathrm{d}K/\mathrm{d}v) \tag{7.11b}$$

The voltage sensitivities are

$$S_T = \frac{1}{V}\left(\frac{\partial V}{\partial T_a}\right) = -\frac{B}{T^2}\left[\frac{1}{1+(BP/KT^2)}\right] \tag{7.12a}$$

$$S_v = \frac{1}{V}\left(\frac{\mathrm{d}V}{\mathrm{d}v}\right) = -\frac{S_T P}{K^2}\left(\frac{\mathrm{d}K}{\mathrm{d}v}\right) \tag{7.12b}$$

For small heating currents the sensitivity to temperature is large and approaches α_a as $I \rightarrow 0$, whereas the sensitivity to velocity variations is small and tends to 0 as $I \rightarrow 0$. As I is increased S_T decreases monotonically and approaches the limit

$$S_T \approx \frac{1}{T} = \frac{K}{(P+KT_a)} \tag{7.13}$$

The sensitivity to velocity increases with I towards the limit

$$S_v \approx \frac{1}{K}\left(\frac{\mathrm{d}K}{\mathrm{d}v}\right) \tag{7.14}$$

For all conditions the ratio of sensitivities is given by

$$\frac{S_v}{S_T} = \frac{P}{K^2}\left(\frac{\mathrm{d}K}{\mathrm{d}v}\right) \tag{7.15}$$

7.2.1.2. CONSTANT TEMPERATURE OPERATION

The discussion so far has assumed constant current. Constant temperature operation can also be used with the thermistor current taken as the output. From Equation 7.10, with T considered a parameter, the constant-temperature sensitivities are defined as

$$I_T = \frac{1}{I}\left(\frac{\partial I}{\partial T_a}\right) = \frac{1}{2(T-T_a)} \tag{7.16a}$$

$$I_v = \frac{1}{I}\left(\frac{\partial I}{\partial v}\right) = \frac{1}{2K}\frac{\mathrm{d}K}{\mathrm{d}v} \tag{7.16b}$$

These may be compared with the constant current sensitivity as follows;

$$\frac{S_T}{I_T} = \frac{S_v}{I_v} = \frac{2B(T-T_a)/T^2}{1+B(T-T_a)/T^2} \tag{7.17}$$

For small power dissipation, constant temperature operation gives greater sensitivities to variation in fluid temperature and velocity; the converse is true for large power dissipation. Equal sensitivities arise for the condition

$$T^2 - B(T-T_a) = 0 \tag{7.18}$$

7.2.1.3. APPLICATION TO INSTRUMENT DESIGN

The theory presented above has been widely implemented. Rasmussen's particular interest was in the design of a bathythermograph, illustrated by the schematic diagram of Fig. 7.3. In this instrument a thermistor, heated by a constant current, is carried down into the depths of the sea by a weight suspended from a light cable. The integrator gives an output voltage proportional to the length of cable and the thermistor voltage drop is amplified to give the necessary electrical signals to operate an X–Y recorder.

If a diving rate of 5 ft/s, a maximum temperature gradient of 0·25 °C/ft and a variational temperature accuracy of 0·05 °C are specified then the design considerations are as follows. From Equation 7.4

$$\frac{\mathrm{d}T}{\mathrm{d}t} = 0{\cdot}15(1-\mathrm{e}^{-t/\tau'}) \tag{7.19}$$

On the assumption that the rate of change attains 99% of the maximum in an interval corresponding to a temperature change of 0·05 deg C an approximate value of the time constant τ' is obtained of 0·014 s. Since the time constant in air is nearly 25 times that in water a thermistor having a time constant in air of approximately 0·355 is required; i.e. a bead-type thermistor is indicated. A standard range of bead-type thermistors has a thermal time constant of about 0·55 s in air;

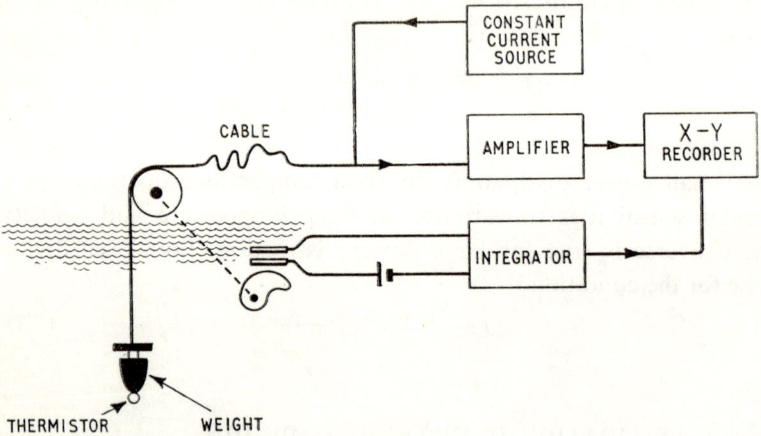

Fig. 7.3. Schematic diagram of a simple bathythermograph (after Rasmussen)

i.e. about 0·02 s in water. This is greater than the value of the effective time constant τ', which is 0·014 s; hence the required reduction can be determined from Equation 7.4 as follows,

$$\frac{0{\cdot}014}{0{\cdot}020} \approx \frac{3}{7} = \frac{BP}{K(T_0+P/K)^2} \qquad (7.20)$$

In solving this equation the maximum expected value of T_0 should be considered. Taking $K = 1{\cdot}45$ mW/K (in water) and $B = 3\,600$ K, with an ambient temperature of 290 K, it follows from Equation 7.17 that the heating power required is $\approx 22{\cdot}5$ mW. This corresponds to a thermistor temperature of about 305·5 K; i.e. $\approx 15{\cdot}5$ K above ambient.

7.2.1.4. OTHER APPLICATIONS BASED ON RATE OF FLOW

Several papers have been written on the application of thermistors to anemometry. These include instrumentation for measuring wind speed[3–7] and forced flow under laboratory conditions. Seymour's anemometer[8], taken as an example, was capable of measuring air speeds between 1 and 300 ft/s, with a claimed accuracy of some 0·5%. This was achieved by having two thermistors—one (already calibrated) in the variable-velocity air stream and the other (a 'calibration' thermistor) in a different part of the system, but at the same temperature, in which the rate of air flow could be controlled to a standard value. By manual operation continual control of the operation point of the 'calibration' thermistor and hence of the 'operating' thermistor

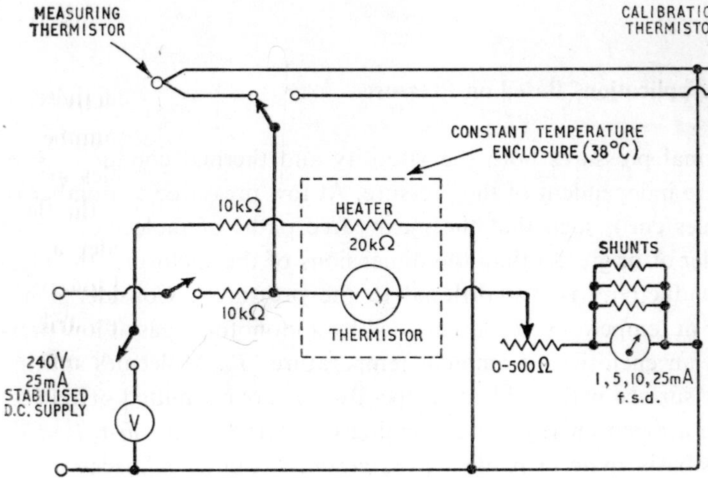

Fig. 7.4. Circuit diagram of Seymour's anemometer

were achieved. The circuit is shown in Fig. 7.4. There are two electrical branches. One contains either the measuring thermistor or the calibration thermistor: the other is comprised of a thermistor in a constant temperature enclosure, a 0–500 Ω variable resistor and a variable range milliammeter reading from 1 mA to 25 mA, full scale deflection. As the air velocity past either the measuring or calibration thermistor increases it is cooled and its resistance rises. This has the effect of increasing the current through the thermistor in the constant temperature enclosure thereby reducing its resistance with consequent further increase in current; the current rise is limited by the series-connected variable resistor, which can be set to give control of sensitivity.

If the range of flow rate to be measured is less extensive at the lower end then a simpler less-sensitive system can be used, in which there is no positive feedback. A common circuit is a Wheatstone bridge with sensing and control thermistors in adjacent arms[9]. The range of measurement is 24–240 ft/s.

Applications to the measurement of moving liquids have been described by several workers[2, 7, 10–13]. References 12 and 13 deal with the flow of blood in humans.

7.2.2. Applications Based on Pressure

At normal pressures both the viscosity and thermal conductivity of gases are independent of the pressure. At low pressure ($\propto$ number of molecules/cm^3), such that the mean free paths of molecules are of the order of or greater than the dimensions of the enclosure, the thermal conductivity is proportional to the pressure[14]. Consider a hot surface at temperature T_s surrounded by a monotonic gas at low pressure in an enclosure at ambient temperature, T_a. Molecules striking the hot surface with incident temperature T_i are re-emitted or reflected with a mean energy corresponding to a temperature $T_i < T_r \leqq T_s$. On this basis an *accommodation coefficient*, α, can be defined;

$$\alpha = \frac{T_r - T_i}{T_s - T_i} \tag{7.21}$$

It is established that in such collisions the mean energy transferred to or from a surface at T, per molecule, is given by $E = 2kT$. Hence if ν represents the number of incident molecules per cm^2 the net energy transfer per second is

$$E_0 = \nu 2k(T_r - T_i)$$

$$= \frac{1}{4} n v_i \, 2k(T_r - T_i)$$

$$= \frac{1}{2} \frac{P v_i}{T_i} (T_r - T_i)$$

$$= \frac{\alpha}{2} \frac{P v_i}{T_i} (T_s - T_i) \qquad (7.22)$$

This follows since a molecule of mass m approaching the surface with incident velocity v_i, rebounding with the same speed will experience a momentum change of $2mv_i$ and if ν molecules of density n per cm^3 strike an area of 1 cm^2 per second the total impulse exerted on the area per unit time is 2 $m\nu v_i$; pressure P, is defined as the rate at which momentum is imparted per cm^2, so that $P = 2\, m\nu v_i$, and it can also be shown that $P = nKT_i$.

The application of heated thermistors to vacuum measurement is obvious. Various instruments have been described[15–26] for measurement in the pressure range $1–10^{-7}$ mmHg, although some do not cover the whole range. That due to Bradley[16] does. It is a modified McLeod gauge, with a bead-type thermistor sealed into the end of the closed capilliary as shown in Fig. 7.5. After calibration the thermistor may be used directly in the range $1–10^{-3}$ mmHg. Pressures in the lower range are measured by compressing the gas in the small chamber

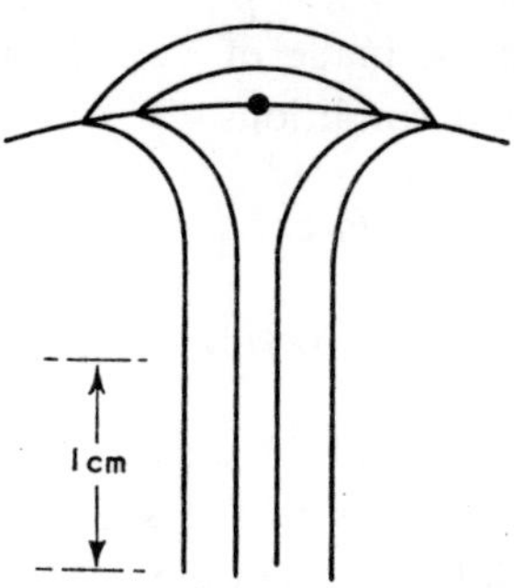

Fig. 7.5. Bradley's McLeod gauge head

enclosing the thermistor until the pressure rises to 10^{-3}–1 mmHg. The pressure in the chamber is then determined and the original very low pressure can be calculated from the compression ratio. Bradley's gauge used a Wheatstone bridge with a balancing thermistor in series with a resistance in an adjacent arm to the sensing thermistor. Varicák and Saftić[21] covered the range 1–10^{-6} mmHg directly by depositing miniature thermistors on thin metal foils.

An associated application which depends on the measurement of change of low pressure with time is in the determination of molecular weight by effusiometry[27]. This involves the release of the gas, whose molecular weight is to be determined, by effusion through a small orifice into a chamber which is initially evacuated. The rate of molecular flow of gas, of molecular weight, M, through a circular opening of radius a in a thin plate is given by[14]

$$F = \frac{\pi a^2 v_a}{4} = 11{\cdot}428 \times 10^3 a^2 (T_a/M)^{1/2} \text{ cm}^3 \text{ s}^{-1} \tag{7.23}$$

where v_a is the average molecular velocity. The volume rate is therefore independent of the pressure. If the gas is flowing into a vacuum from a vessel with a fixed volume, V_0, and at an initial pressure, P_0, then the instantaneous rate of flow is an invariant fraction of the total volume, and the pressure, p, of the gas in the initially evacuated chamber at a given time, t, is

$$p = P_0 \exp\left\{\left(-\frac{a^2}{V_0}\right)\left(\frac{RT\pi}{2M}\right)\right\}^{1/2} t \tag{7.24}$$

Hence, if a comparison is made of various gases effusing from the same vessel under similar conditions

$$p = P_0 \exp\left(-\frac{kt}{M^{1/2}}\right) \tag{7.25}$$

or

$$\log_e p = -(kt/M^{1/2}) + \log_e P_0 \tag{7.26}$$

Thus a semi-logarithmic plot of the pressure of the effusing gas against time has a slope which varies as the square root of the molecular weight. A convenient datum is the time $t_{1/2}$, required for the pressure to drop to half its value at some previous instant. A typical result for hydrogen at two different temperatures is shown in Fig. 7.6. If the size of the hole is measured accurately then the effusion rate for any gas can be calculated. If this is not practical then it is necessary to calibrate the instrument using known gases. Such a calibration, in terms of $t_{1/2}$ as defined above is shown in Fig. 7.7.

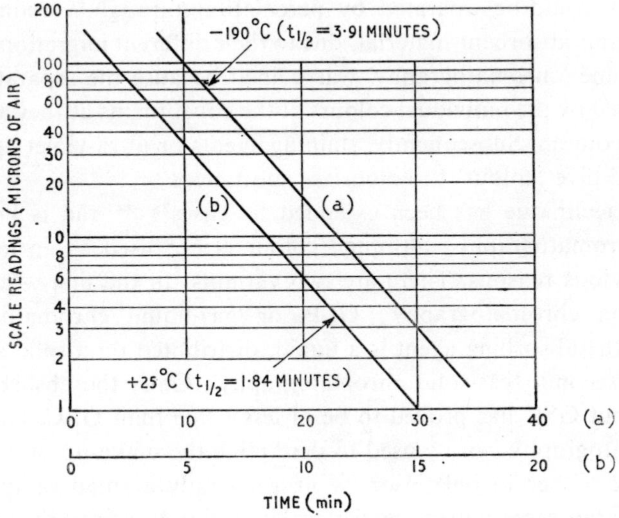

Fig. 7.6. Effusion rate of hydrogen at; (a) 25 °C, (b) – 190 °C (after Zemany)

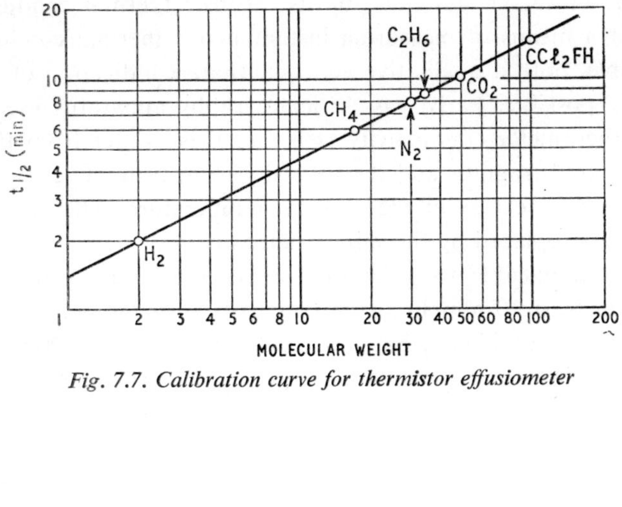

Fig. 7.7. Calibration curve for thermistor effusiometer

7.2.3. Applications Based on Different Thermal Conductivities of Individual Gases and Vapours

Chromatography emerged as a chemical analytical technique from the discovery in 1903 by Tswett that the constituents of vegetable extracts could be separated by percolation through a column containing an absorbent material, due to their different migrational rates. The name 'chromatography' was coined because the separation was indicated by the individual colours of the constituents as they appeared in the column. Subsequently, staining agents or ultra-violet light were used to give 'colour' to colourless substances.

The technique has been extended to gases[28–33] and is known as 'gas-chromatography', although colour is not used as an indicator, for obvious reasons. There are two variants. In the first—'gas-liquid partition chromatography', GLP, or 'gas-liquid chromatography', GLC, the absorbing agent is a liquid, distributed on a solid support: in the second, 'gas-solid chromatography', GSC, the absorbent is a solid, but GSC has proved to be of less value than GLC. Gas-liquid chromatography can be used to determine the make-up of a gaseous mixture, either in bulk,[31] or by utilising only a small sample or it. In the latter case an inert 'carrier gas' is used to transport the unknown sample into the column and through it. The emergence from the column of a particular constituent of a gaseous mixture is sensed, for a fixed rate of carrier gas flow, by its intrinsic thermal conductivity. Hence, if a means of measuring instantaneous thermal conductivity is available, both a qualitative and quantitative indication of constituents is possible. Early gas-chromatography apparatus used 'hot-wire' katherometer conductivity cells, but these suffer from the disadvantages of fragility and the necessity to operate at temperatures considerably in excess of the ambient temperature. They have now been largely replaced by thermistor detectors.

GLC is applicable to gases or samples that exert vapour pressures of at least 10 mmHg at the temperature of the column, which can be 300 °C or higher, but not greater than about 1000 mmHg, for which 'absorption chromatography' is more effective.[30]

There is now a considerable literature on gas chromatography. The major applications are based on elution* analysis[30]. In this the column is filled with a solid absorbent such as kieselguhr impregnated with an inert solvent such as stearic acid.

The sensing head for GLC is a thermal-conductivity cell. This is shown schematically in Fig. 7.8.[32, 33] It consists of a metallic block,

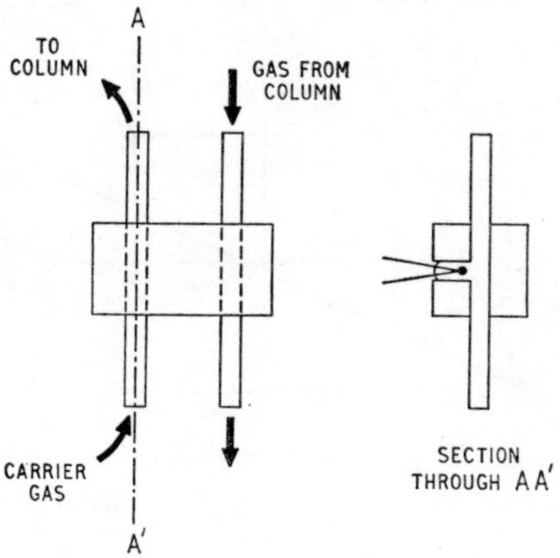

Fig. 7.8. Schematic diagram of thermal conductivity cell

of large heat capacity, through which both the input and output gaseous streams from the column flow. Holes bored transversely to the direction of 'gas flow' with appropriate seals, contain matched thermistors, which are connected electrically in the usual Wheatstone bridge circuit. In view of the sensitivity of thermistor resistance to temperature and gas flow rate, both of these have to be controlled.

Typical calibration curves[31] for binary gas systems, relating bridge output to gas composition for the purpose of bulk gas analysis are shown in Fig. 7.9. In Fig. 7.10 a gas-chromatogram is shown[33] for a system in which helium gas is used as the carrier and the benzene sample investigated contained 70 parts per million (p.p.m) isopropyl alcohol. In this particular case, which is typical, 1% of full scale deflection represents 1 p.p.m. of isoprophyl alcohol. By increasing the

* Elution is the opposite of absorption.

amplification of the bridge output or increasing the sample size, less than 1 p.p.m. of contaminant can be detected. In the chromatogram shown the other peaks were not evaluated. The peaks and associated skirts as shown in Fig. 7.10 are referred to as 'elution bands'. With suitable calibration the areas of these bands can be used for quantitative analysis.[30]

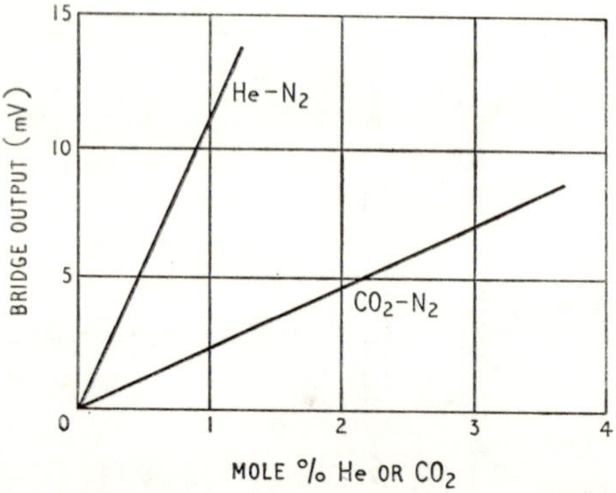

Fig. 7.9. Calibration curves for gas-chromatography (after Walker and Westenberg)

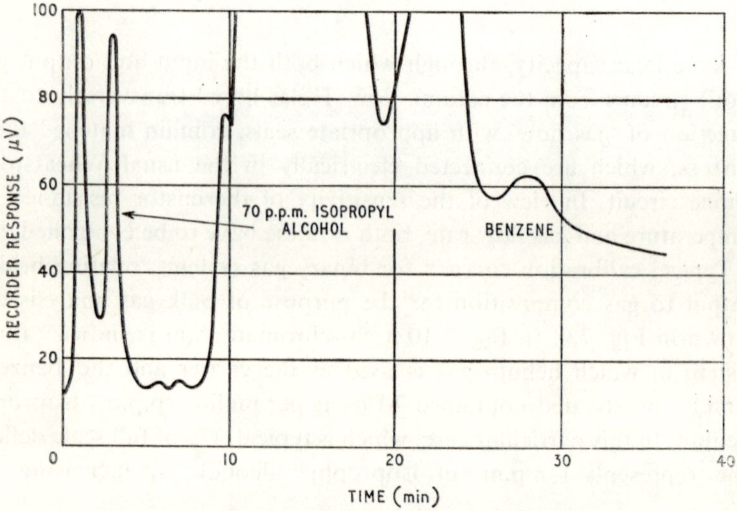

Fig. 7.10. Typical gas-chromatogram (after Bennett et al)

7.3. LIQUID-LEVEL SENSING

In cryogenic systems it is common to use a temperature-sensitive resistive element to indicate the cryogenic liquid level. This is indicated by the difference between heat transfer from the liquid and vapour phases and the sensing element. Initially carbon resistors were commonly used but it was found that continual cycling between room temperature and cryogenic temperatures tended to produce permanent changes in both the resistance and temperature coefficient. At liquid nitrogen temperature the liquid to vapour resistance change, (typically about 30%) was great enough to mask these changes, but at liquid helium temperature, with a resistance change of about 5%, frequent calibration was often necessary.

N.T.C. thermistors have been used successfully without showing this behaviour and with advantage because of their greater temperature coefficient of resistance. Lewin[34] used S.T.C. type *E* 23 thermistors, having a room temperature resistance of about 2 kΩ, but a resistance at liquid helium temperature in excess of 10^9 Ω. In view of the difficulty of obtaining a satisfactory operating point by self-heating he wound a 44 s.w.g. manganin heater coil round the thermistor body. The dissipation of 5 W in the heater raised the thermistor temperature from 4·2 K to room temperature in under 2 s when in helium vapour, but barely changed its temperature when in liquid helium. In view of the difference in room temperature resistance of 2 kΩ and the liquid helium temperature resistance of $\approx 10^9\,\Omega$ the thermistor is an effective switch in an electrical circuit. Salomaa[35] used self-heated Fenwal GD 2382 thermistors in the design of an automatic liquid nitrogen pumping system which found widespread use. Hogan et al[36] have drawn attention to the effect of contact resistance in some thermistors with low resistance at cryogenic temperatures. High resistance thermistors do not appear to be affected and their d.c. characteristics are linear.

7.4. DETAILED DESIGN CONSIDERATIONS

The approach to design is best made graphically if operation over an extensive range of the characteristics is involved, but over small ranges a small-signal approach can be used. Both of these approaches will be considered. The emphasis here will be on practical circuit design for instrumentational purposes, rather than on general principles of operation, which have been emphasised in other chapters. The three designs considered follow a paper by Sapoff and Oppenheim[1].

7.4.1. Graphical Approach

There are three basic ways in which a thermistor can be used to give an output indicative of a change in K. These are: in a simple potential divider circuit; in a Wheatstone bridge in which the other arms are passive resistors; and in a Wheatstone bridge incorporating a second thermistor in an adjacent arm, whose function is to nullify the effects of any ambient temperature changes. These are in the direction of increasing complexity.

7.4.1.1. SIMPLE POTENTIAL DIVIDER CIRCUIT

The circuit to be considered is shown in Fig. 7.11. The output voltage E_0 is taken across R. The determination of output against input change is straightforward. A family of static current-voltage characteristics is first drawn, covering the range of dissipation constant to be encountered. On these the load line corresponding to E and R is then placed. The choice of E and R is governed by the requirements for stability. Clearly a load line which is close to being tangential to the static characteristic through the operating point will give the maximum output indication, but this would be an unsuitable choice because slight changes in temperature could create instability. A practical solution is to set R equal to 1·5 to 2·5 times the slope resistance of the characteristic at the operating point. This is illustrated in Fig. 7.12a and the corresponding dynamic characteristic relating E_0 to K in Fig. 7.12b. The operating point should be chosen as close to the peak of the characteristic as possible because here individual curves are most widely separated.

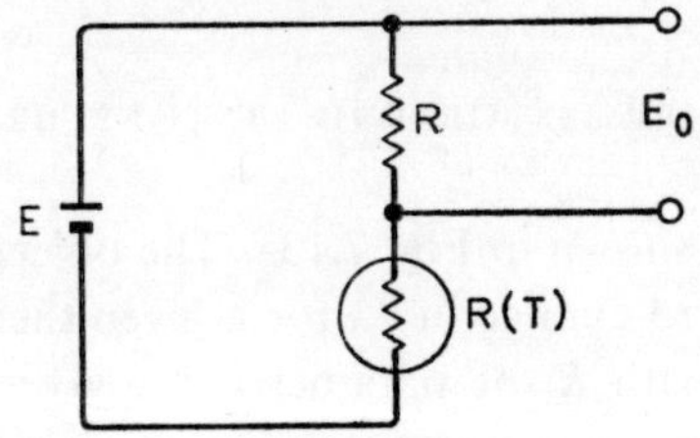

Fig. 7.11. Simple potential divider circuit

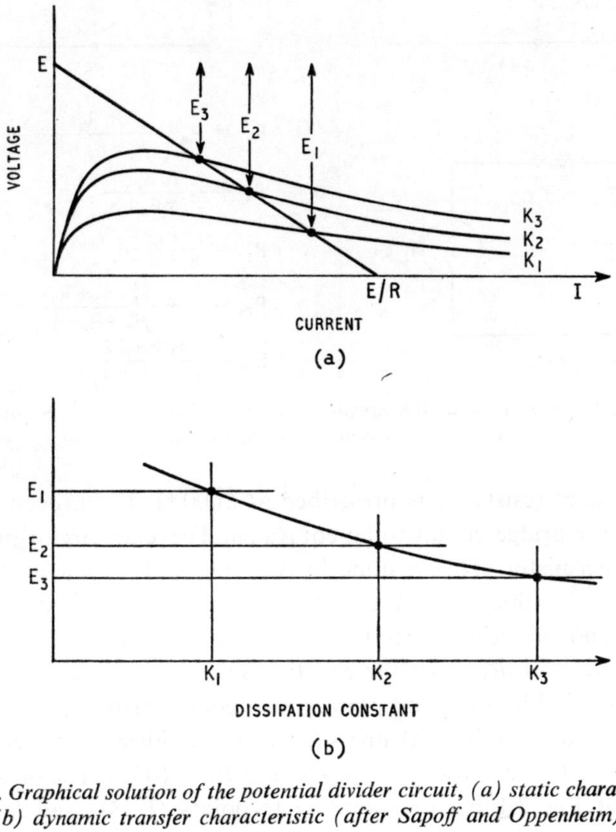

Fig. 7.12. Graphical solution of the potential divider circuit, (a) static characteristic; (b) dynamic transfer characteristic (after Sapoff and Oppenheim)

7.4.1.2. SINGLE-THERMISTOR WHEATSTONE BRIDGE

The basic circuit is shown in Fig. 7.13a. The objective is optimisation of output response to change of K, for a given thermistor, whose static characteristics, with K the parameter, are given in Fig. 7.14. This thermistor has a zero power room temperature resistance of 8000 Ω

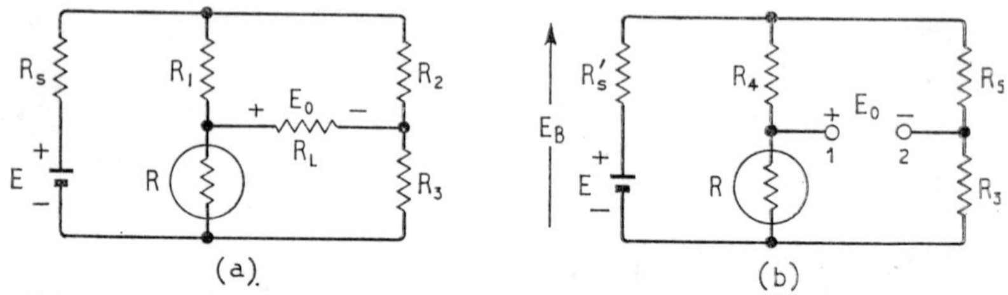

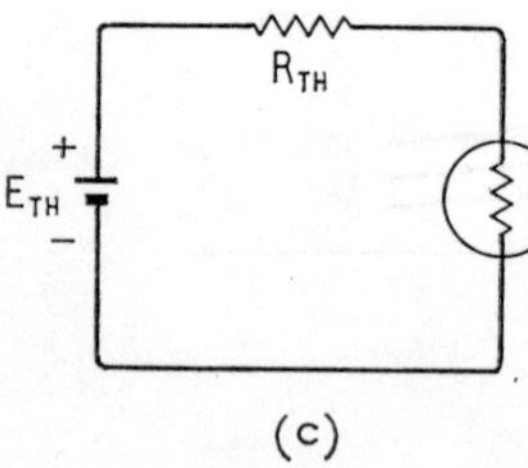

$$R_S' = R_S + \frac{R_1 R_2}{R_1 + R_2 + R_L}$$

$$R_4 = \frac{R_1 R_L}{R_1 + R_2 + R_L}$$

$$R_5 = \frac{R_2 R_L}{R_1 + R_2 + R_L}$$

$$R_{TH} = R_4 + \frac{R_S'(R_3 + R_5)}{R_S' + R_3 + R_5}$$

$$E_{TH} = \frac{(R_3 + R_5)E}{R_S' + R_3 + R_5}$$

Fig. 7.13. Single-thermistor Wheatstone bridge; (a) basic circuit; (b) transformed circuit; (c) reduced circuit (after Sapoff and Oppenheim)

and the load resistance is prescribed as 2000 Ω. The procedure is to convert the bridge circuit to an equivalent Thévenin circuit presented to the thermistor. This is done in two stages. The first is shown in Fig. 7.13b in which a Δ–Y transformation has been used on R_1, R_2 and R_L and this leads directly to the equivalent circuit of Fig. 7.13c where E_{TH} and R_{TH} are the effective source e.m.f. and load resistance. A suitable operating point giving good sensitivity and stability is M_1 and a suitable load line intersects the other characteristics at Q_1 and N_1. This defines $E_{TH} = 7{\cdot}4$ V and $R_T = 643\,\Omega$. The bridge voltage E_B of Fig. 7.13b corresponding to points Q_1, M_1 and N_1 can be found by drawing a line of slope $-R_4$ through each of them. A value of R_4 may be derived using the following argument.

It is to be noted that as R_s' goes to zero R_4 approaches R_{TH} and the variation in bridge voltage E_B with changing thermistor resistance goes to zero and this means that the voltage across R_3 does not vary.

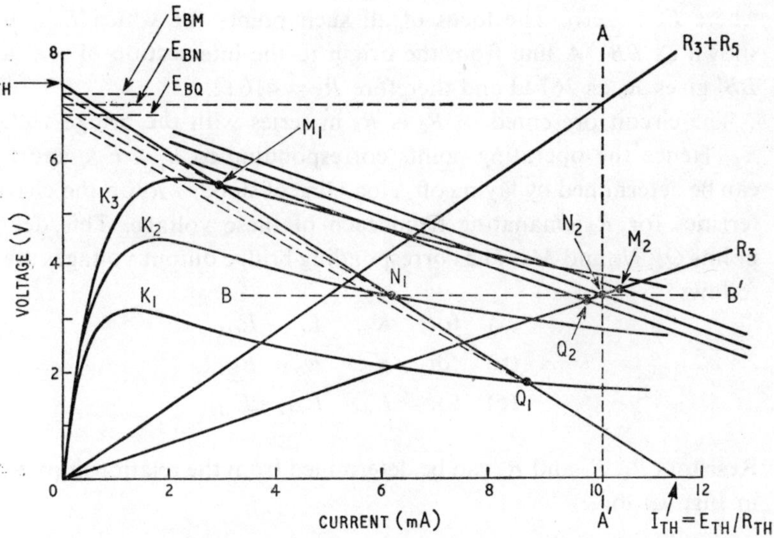

Fig. 7.14. Static characteristics of thermistor type VECO 38C2 with dissipation constant as parameter (after Sapoff and Oppenheim)

This is the condition for maximum sensitivity. It cannot be reached but a reasonable compromise is

$$(R_3+R_5)\,||\,R_s'+R_4 = R_{TH} = 1{\cdot}1R_4 = 643\,\Omega$$

giving $R_4 = 585\,\Omega$ and $(R_3+R_5)\,||\,R_s' = 58\,\Omega$. Using the value of R_4 so derived gives the corresponding values E_{BQ}, E_{BN} and E_{BM} shown in Fig. 7.14. For later convenience it is found that E should be chosen slightly higher than E_{TH} and in this case a standard e.m.f. is $E = 8{\cdot}1$ V. This gives

$$E_{TH} = \frac{(R_3+R_5)}{(R_s'+R_3+R_5)}\,8{\cdot}1 = 7{\cdot}4\text{ V}$$

and $\dfrac{R_3+R_5+R_s'}{R_3+R_5} = 1+\dfrac{R_s'}{R_3+R_5} = \dfrac{E}{E_{TH}} = 1{\cdot}093$ so that $R_s'/(R_3+R_5) = 0{\cdot}093$ and $R_s' = 1{\cdot}093\times 58 = 63{\cdot}5\ \Omega$, and $R_3+R_5 = 683\ \Omega$.

The next stage is the determination of R_3 and R_5. This is done by laying off a line of slope R_3+R_5 from the origin. The line AA' corresponds to the current flowing through R_3+R_5 at the operating point N_1 where E_0 is zero. The locus of all such points for which $E_0 = 0$ is shown as BB'. A line from the origin to the intersection of AA' and BB' gives R_3 as 267 Ω and therefore $R_5 = 416\ \Omega$.

The circuit presented to R_3 is R_5 in series with the bridge voltage E_B. Hence the operating points corresponding to E_{BQ}, E_{BN} and E_{BM} can be determined by laying off a load line of slope $-R_5$ on the characteristics for R_3 emanating from each of these voltage. This defines points Q_2, N_2 and M_2. The corresponding bridge output voltages are as follows:

(a) for K_1, $E_{Q2}-E_{Q1}$

(b) for K_2, $E_{N2}-E_{N1}$

(c) for K_3, $E_{M2}-E_{M1}$

Resistors R_1, R_2 and R_3 can be determined from the relationships given in Fig. 7.13b.

7.4.1.3. DOUBLE WHEATSTONE BRIDGE

The third detailed example given by Sapoff and Oppenheim is the design of a thermal conductivity gas analyser. The specific application chosen is to the determination of the average molecular weight of an unknown gas from the binary systems shown in Fig. 7.15 using air as a reference gas with both this and the unknown gas maintained at zero flow and 25 °C.

The proposed circuit is shown in Fig. 7.16a, incorporating matched thermistors to give a balance when both are in air. The VECO thermistors chosen (type A 140) have a normal resistance of 8000 Ω at 25 °C. The data presented in Fig. 7.15 is obtained with the thermistors mounted in cell blocks identical to those to be used in the intended application—a necessary precaution because of the strong dependence of dissipation constant on the mounting and environment. The limit points imposed correspond to air with a molecular weight of about

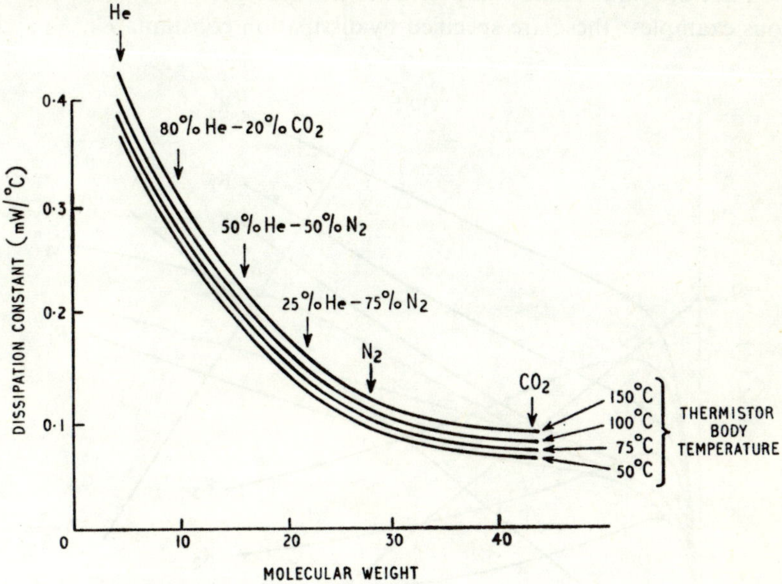

Fig. 7.15. Dependence of thermistor dissipation constant on molecular weight of a binary gaseous mixture (after Sapoff and Oppenheim)

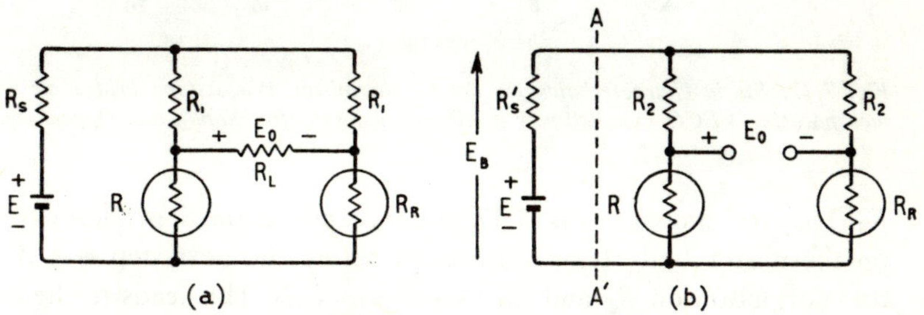

$$R_2 = \frac{R_1 R_L}{2R_1 + R_L} \qquad R_s' = R_s + \frac{R_1^2}{2R_1 + R_L}$$

Fig. 7.16. Double thermistor Wheatstone bridge; (a) basic circuit, (b) transformed circuit (after Sapoff and Oppenheim)

28·8 and helium with a molecular weight of about 4, with an approximate midpoint condition corresponding to 50% He and 50% N_2 (molecular weight $\approx$ 16).

First the three static characteristics are synthesised as in the previous examples: these are specified by dissipation constants K_1, K_2 and

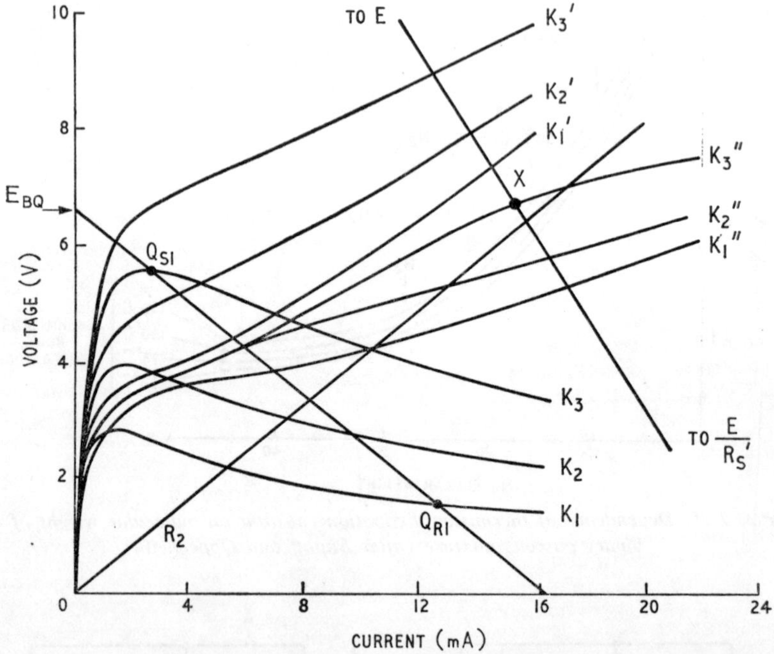

Fig. 7.17. Static characteristics for double thermistor Wheatstone bridge circuit, incorporating VECO matched pair A140 thermistors (after Sapoff and Oppenheim)

K_3. They are shown in Fig. 7.17, and incorporate the dependence of K on thermistor body temperature. As before the next step is a Δ–Y transformation on R_L and the two resistors R_1. This leads to the circuit of Fig. 7.16b.

Again it is a reasonable compromise to choose values of R_2 some 1·5 to 2·5 times the magnitude of the slope of the static characteristic at the working point. This is of the order of 200 Ω, so 400 Ω is chosen for R_2. The location of the load line is chosen to indicate maximum output when the gas being sensed is He. This is Q_{S1}, the reference thermistor in air being operated at Q_{R1}. The bridge voltage E_{BQ} is given by the intersection of the load line for R_2 on the voltage axis. It is evident that the maximum output voltage from the bridge is the difference between the voltages corresponding to Q_{S1} and Q_{R1}; i.e. 4 V. The output detector is chosen as a 1 mA full scale meter. For full-scale deflection with 4 V R_L is 4000 Ω. With R_L and R_2 known R_1 is found from $R_2 = R_1R_L/(2R_1+R_L)$ to be 500 Ω. Since $R_s' = R_s + R_1^2/(2R_1+R_L)$ this means that $R_s = R_s' - 50$ Ω. Denoting the combined current corresponding to Q_{S1} and Q_{R1} as I_T we may write

$$E - I_T R_s' = E_{BQ}$$

or

$$E - 15{\cdot}4 \times 10^{-3} R_S = 7{\cdot}45\ V \qquad (7.27)$$

Any pair of values of E and R_S satisfying this equation will produce the required operating conditions.

To be able to predict the molecular weight of a gas from the out-of balance voltage reading, appropriate combined characteristics for the circuit to the right of AA' in Fig. 7.16b must be drawn. This is done in two stages. First the characteristics K_1', K_2' and K_3' for the series combination of thermistors and R_2 (ohmic line from the origin) is obtained by adding ordinates. Then by adding current values at each voltage the final set K_1'', K_2'', K_3'' is obtained. On these characteristics the load line corresponding to R_s' and E is now laid off. The constraint of Equation 7.27 requires that this should pass through the point X on curve K_3'' with the voltage coordinate E_{BQ}. The values of E_B corresponding to any other operating point can now be read off from the points of intersection of the R_s' load line and the K'' family of curves.

The corresponding operating points on the K family of characteristics can now be obtained from the intersection of a load line of slope $-R_2$ with them, for an effective source voltage equal to the appropriate value of E_B.

A comparison of predicted bridge output and measured bridge output for gaseous mixtures having a range of molecular weights is shown in Fig. 7.18.

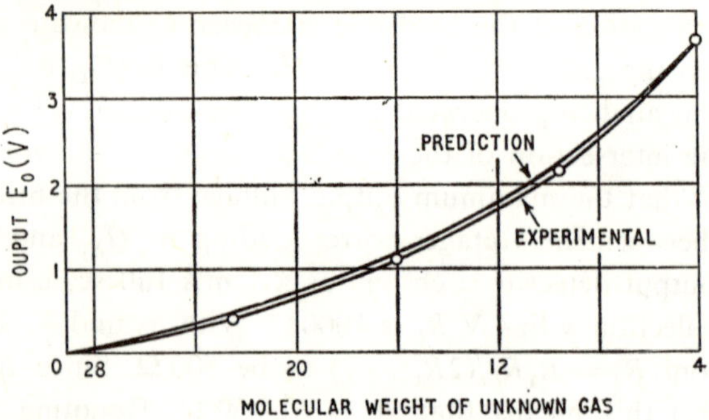

Fig. 7.18. Comparison of predicted and measured molecular weights of binary gas mixtures (after Sapoff and Oppenheim)

REFERENCES

1. SAPOFF, M., and OPPENHEIM, R. M. 'Theory and application of self-heated thermistors', *Proc. Inst. elect. electron. Engrs.*, **51,** 1292 (1963)
2. RASMUSSEN, R. A. 'Application of thermistors to measurements in moving fluids', *Rev. Scient. Instrum.* **33,** 38 (1962)
3. EYRING, C. F. 'Jungle acoustics', *J. acoust. Soc. Am.*, **18,** 257 (1946)
4. HALES, W. B. 'Thermistors as instruments in thermometry and anemometry', *Bull. Am. met. Soc.*, **29,** 494 (1948)
5. CHUDNOVSKI, A. F. 'The use of semiconductors in investigating natural phenomena', *Zh. Tekh. Fiz.*, **25,** 2122 (1955)
6. POLLACK, H. 'An electronic anemometer', *Pop. Electron.*, **6,** 51 (1957)
7. VEPŘEK, J. 'Analysis of the characteristics of thermistors as the elements for measuring the velocity of liquids and gases', *Slabopr. Obz.*, **20,** 353 (1959)
8. SEYMOUR, E. V. 'Simple miniature probe anemometer' *Engineer, London,* **215,** 899 (1963)
9. HOWES, J. W. 'Characteristics of thermally-sensitive resistors' *Proc. Instn. Radio Engrs Aust.*, **13,** 123 (1952)
10. VEPŘEK, J. 'The measurement of very low velocities of liquids by means of thermistors', *Slabopr. Obz.*, **19,** 585 (1958)
11. REVERSEAU, J. 'L'Emploi des thermistances pour la mesure des très faibles vitesses dans les liquides', *Génie Civ. (Hydraulique)*, **136,** No. 18, 365 (1959)
12. COOPER, R. 'Local changes in intra-cerebral blood flow and oxygen in humans', *Med. Elect. and Biol. Engng.* **1,** 529 (1963)
13. JAMES, G. W., PAUL, M. H., and WESSEL, H. U. 'Thermal dilution: instrumentation with thermistors', *J. Appl. Physiol.* **20,** 547 (1965)
14. DUSHMAN, S. *Scientific foundations of vacuum technique*, John Wiley (1949)
15. DUSHMAN, S. 'Manometers for low pressures', *Instruments*, **20,** 234 (1947)

16. BRADLEY, R. S. 'A thermistor McLeod gauge for a pressure range 1–10^{-7} mm of mercury', *J. Scient. Instrum.*, **31,** 129 (1954)
17. VARIČAK, M. 'L'influence de la température de l'enciente sur la sensibilité du manomètre thermique', *J. Phys. Rad. Phys. Appliqués*, **18,** 70A (1957)
18. VARIČAK, M. 'Experiments in the use of thermistors for low pressure measurements', *Nuovo Cim.*, **6,** 723 (1957)
19. KORNER, R. L. and RIEGER, G. N. 'Thermistor checks tube exhausting', *Electron. Inds Tele–Tech* 1957, **16,** 60, 153
20. DAVIS, F. E. 'Vacuum furnace control', *Electronics*, **21,** 81 (1958)
21. VARIČAK, M. and SAFTIC, B. 'Principle of a semiconductor manometer in the pressure range of 1 to 10^{-6} mmHg *Rev. Scient. Instrum.*, **30,** 891 (1959)
22. FLANICK, A. P. and AINSWORTH, J. E. 'Thermistor pressure gauge design', *Rev. Scient. Instrum.*, **32,** 356 (1961)
23. HANSEN, N. and SCHMIDT, W. 'Applications of N.T.C. thermistors in vacuum measurements', *Matronics*, **20,** 400 (1962)
24. VEPŘEK, J. and ZOBAČ, L. 'A thermistor vacuum gauge', *Slabopr. Obz.*, **25,** 34 (1964)
25. ROBERTS, R. W., MCELLIGOT, P. E. and JERKANOFF, G. 'A bakeable thermistor vacuum gauge', *J. Vac. Sci. Techn.*, **1,** 62 (1964)
26. BERRY, C. J. 'Low pressure measurement with thermistors', *J. Scient. Instrum.*, **44,** 83 (1967)
27. ZEMANY, P. D. 'Molecular weight determinations by effusiometry', *Rev. Scient. Instrum.*, **23,** 176 (1952)
28. JAMES, A. T. and MARTIN, A. J. P. 'Gas Liquid Partitition Chromatography. A Technique for the analysis of Volatile Materials', *Analyst*, **77,** 915 (1952)
29. DESTY, D. H. (ed.) *Vapour phase chromatography*, (paper by Hansdorff and discussion by Minkoff) Academic Press Inc. New York (1957)
30. LATINEN, H. A. *Chemical analysis* McGraw-Hill Book Co. Inc. (1960)
31. WALKER, R. E. and WESTENBERG, A. A. 'Precision thermal-conductivity gas analyser using thermistors', *Rev. Scient. Instrum.* **28,** 789 (1957)
32. AMBROSE, D. and COLLERSON, R. R. 'A thermal conductivity gauge for use in gas-liquid partition chromatography', *J. Scient. Instrum.*, **32,** 323 (1955)
33. BENNETT, C. E., DAL NOGARE, S., SAFRANSKI, L. W. and LEWIS, C. D. 'Trace analyses by gas chromatography' *Analyt. Chem.*, **30,** 898 (1958)
34. LEWIN, J. D. 'A cryogenic liquid level detector employing thermistors', *J. Scient. Instrum.*, **40,** 539 (1963)
35. SALOMAA, M. K. 'Automatic liquid nitrogen pumping system for particle accelerators'. *Nucl. Insrum. Meth.*, **24,** 477 (1963)
36. HOGAN, W., NICOL, J., and SHAPIRO, S. 'A comment on the characteristic of thermistors at 4·2 K; *Adv. cryogen. Engng.*, **8,** 340 (1963)

CHAPTER 8

Applications Based on Resistance-Temperature Characteristics

8.1. INTRODUCTION

One of the most extensive fields of application of thermistors is in temperature measurement and sensing, and also temperature control and alarm systems. They can be used as well to compensate for changes in resistance of circuit components having a temperature-coefficient of resistance of opposite sign: N.T.C. thermistors, in conjunction with passive linearising resistors, are mainly used for this purpose to compensate for the positive increase of resistance with temperature of metallic conductors. The features of thermistors which lend them to these applications are:

(a) a large temperature coefficient of resistance
(b) small size
(c) the ability to withstand electrical and mechanical stresses
(d) the ability to operate over a wide range of temperature
(e) the large range of resistance values which can be realised.

In this context it should be noted that perovskite-type P.T.C. thermistors would not normally be used for temperature measurement, because of the additional contact non-linearity, which is inherent.

For temperature measurement the non-linear resistance-temperature characteristic of N.T.C. thermistors is a disadvantage, but techniques are available for changing this to a linear one. The variation of α with temperature for S.T.C. material is given in Table 8.1.[1]

Table 8.1. VARIATION OF TEMPERATURE COEFFICIENT WITH TEMPERATURE

Absolute temperature K	*Resistance* kΩ	*Temperature coefficient* %
200	0·15	6·7
273	4·1	3·6
300	1·7	3·0
373	0·25	1·9
473	0·034	1·2

8.2. TEMPERATURE MEASUREMENT

In general a thermistor is chosen for temperature measurement where remote indication is required, small size is essential, ruggedness is necessary or where a small temperature difference is involved. A platinum resistance thermistor would be chosen where long-term stability is of highest importance or where the temperature is too high for a thermistor (> 350 K). Thermocouples are only useful for measurement of temperature differences and have lower sensitivity.

For a simple thermistor system it is sufficient to monitor the current through a series combination of a passive resistor and a thermistor, fed from a constant voltage supply. However, the majority of applications involve the use of a Wheatstone bridge to eliminate the standing current, with the thermistor or preferably the thermistor and compensating network occupying one arm; the other arms ideally are temperature-insensitive resistors.

Direct-reading thermistor bridges give an indication of temperature either in terms of the unbalance current in the detector or a calibrated resistance arm, which is adjusted to give a balance at a fixed temperature.

The other type of bridge, which is useful for many applications, such as thermometric chemical analysis and dew-point hygrometry, is designed to measure temperature differences.

Common to both types of bridge is the problem of decrease of rate of change of (N.T.C.) thermistor resistance with increasing temperature. This leads to a temperature-dependent sensitivity. A second problem is the spread of constants for thermistors, even of a given type.

8.2.1. Direct-reading Bridges

Early bridges have been described by Andrews[2], Deeter[3], Herington and Handley[4], Greenhill and Whitehead[5] and Macradyen[6]. No specific attempt was made to obtain a linear thermometer, although Macradyen stated that 'by empirical methods combinations of resistors and thermistors may be found by means of which any logarithmic law may be simulated.'

8.2.1.1. LINEARISATION

Beakley[7] considered analytically the linearisation of the inherently non-linear dependence of current and voltage of an N.T.C. thermistor in series with a resistance r and e.m.f. E. He assumedthat the thermistor resistance R was given by Equation 2.7; i.e. $R = R_c \exp(B/T)$. The current I is given by

$$I = \frac{E}{R+r} \tag{8.1}$$

He expanded the current into an infinite series about the working temperature T_0 as follows:

$$I(T) = I(T_0)+hI'(T_0)+\frac{h^2I''(T_0)}{2!}+\ldots+\frac{h^nI^n(T_0)}{n!} \tag{8.2}$$

where $h = T - T_0$ and I', I'' ... I^n are the first, second and nth derivatives of I with respect to T. His approach to linearisation was based on making the second derivative on the right hand side of Equation 8.2 zero. The condition, obtained by differentiating Equation 8.1 twice with respect to T is

$$r = \frac{B-2T_0}{B+2T_0} R_0 \tag{8.3}$$

where $R_0 = R_c \exp(B/T_0)$. On substitution of this value for r into higher derivatives in Equation 8.2 this becomes

$$I(T) = I_0 + hI'(T_0) - \frac{h^3 B^2}{3!\,2T_0^4} I'(T_0) + \frac{h^4 2B^2}{4!\,T_0^5} I'(T_0) \tag{8.4}$$

The terms in h^3, h^4 etc indicate departure from linearity and may be regarded as error terms, ε_3, ε_4 etc. ε_4 is negligible compared with ε_3 for small values of h, since

$$\frac{\varepsilon_4}{\varepsilon_3} = -\frac{2h^4 B^2}{4!\,T_0^5} I'(T_0) \bigg/ \frac{h^3 B^2}{3!\,2T_0^4} I'(T_0) = -\frac{h}{T_0} \tag{8.5}$$

This shows, for example, that at all values of $T_0 > 200$ K $\varepsilon_4 < 0{\cdot}1\ \varepsilon_3$, if $h < 20$ K. Hence only ε_3 need be considered. On this basis Beakley showed that if it is required to construct a thermometer to measure temperature from 290 K to 310 K with a thermistor for which $B =$ 3000 K, the departure from linearity has a maximum value of 0·03 °C. If an error of 0·1 °C is permissible then a range from 285 K to 315 K may be covered.

8.2.1.2. INCORPORATION OF THE THERMISTOR IN A WHEATSTONE BRIDGE

It will be shown that when a thermistor is incorporated in a Wheatstone bridge in which the other components are passive, the expression for the current through it is the same as Equation 8.1, with the simple modification of a constant multiplier. Hence, the previous analysis for a simple series circuit is also valid for a bridge circuit. The circuit to be considered is shown in Fig. 8.1.

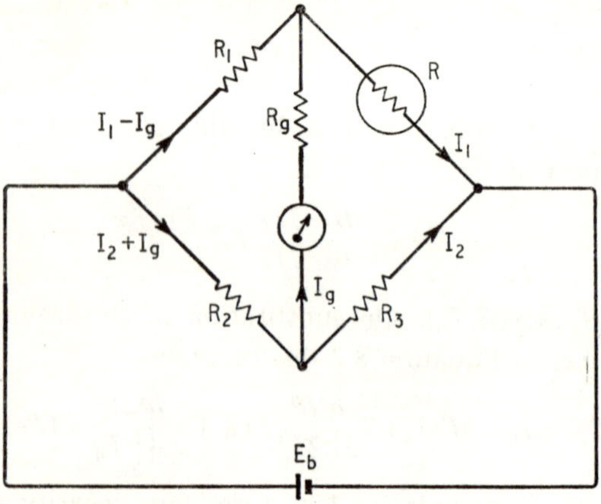

Fig. 8.1. Wheatstone bridge circuit for temperature measurement

The three equations defining this network are:

$$\left.\begin{aligned} E_b &= R_1(I_1-I_g)+RI_1 \\ E_b &= (R_2+R_3)I_2+R_2I_g \\ O &= RI_1-I_2R_3+R_gI_g \end{aligned}\right\} \tag{8.6}$$

Elimination of I_1 and I_2 gives

$$I_g = \frac{E_b(R_1R_3-R_2R)}{R(R_1R_2+R_1R_3+R_2R_3+R_2R_g+R_3R_g)+R_1(R_2R_3+R_2R_g+R_3R_g)} \tag{8.7}$$

which may be rewritten as

$$I_g = \frac{E_b(l-mR)}{pR+q} \tag{8.8}$$

where l, m, p, q are constants. An alternative form is

$$I_g+\text{constant} = \frac{E_b(l/p+mq/p^2)}{R+q/p}$$

This can be compared with Equation 8.1 and may be recast as

$$I = \frac{E_bs}{R+r} \tag{8.9}$$

where s is a constant which may be termed a sensitivity-reduction factor;

$$s = \left(\frac{l}{p} + \frac{mq}{p^2}\right) \tag{8.10}$$

Equations 8.1 and 8.9 are of the same form in R. Hence Equation 8.9 is linear in T for the same conditions that Equation 8.1 is linear in T; i.e.

$$r = \frac{q}{p} = \frac{r_1[r_g(r_2+r_3)+r_2r_3]}{(r_1+r_g)(r_2+r_3)+r_2r_3} = \frac{r_1x}{r+x} \tag{8.11}$$

The relationship between galvanometer current and temperature can be shown to be

$$\frac{dI_g}{dT} = \frac{sE_b(B+2T_0)^2}{4T_0^2BR_0}\left(1-\frac{h_0^2B^2}{12T_0^4}\right) \tag{8.12}$$

where $\pm h_0$ represents the limits of the range of the thermometer.

In the design of this direct-reading thermometer r is calculated from Equation 8.3 and s from Equation 8.10.

If a current-limiting resistor is used in series with the battery supplying the bridge the expression for galvanometer current again is of the form of Equation 8.1 and a linear response is approximated.

It is implicit in the above discussion that self-heating of the thermistor is insignificant. The effect of small self-heating is considered by Beakley.

Mclean[8] relaxed Beakley's design criteria to some extent and produced a skin-thermometer with 'linear' calibration in the range 25–45 °C, with an accuracy of 0·1 °C over this range. Cole has also considered linearity and sensitivity for a range of temperatures.[9]

8.2.1.3. CONSTANT SENSITIVITY

Since the rate of change of (N.T.C.) thermistor resistance with temperature decreases with temperature, the sensitivity of the simple bridge considered so far decreases as the working temperature increases. Pitts and Priestley[10] have described a bridge in which near-constant sensitivity can be achieved over a limited range of temperature.

Their circuit is shown in Fig. 8.2. The novel feature is a tapped resistance A, part of which is in series with the thermistor and the remainder in series with the supply to the bridge. With this arrangement it has been shown, to a first approximation, that dI_g/dT is independent of R and hence of T if $2R_aR_b/(R_a+R_b) \approx R_b-A$.

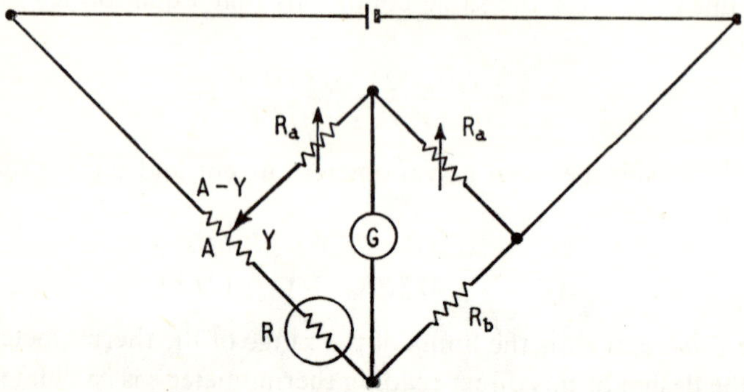

Fig. 8.2. Thermistor bridge giving a near-constant sensitivity over a limited range of temperature (due to Pitts and Priestley)

A thermometer based on this design, checked at various temperatures against a Beckmann thermometer, gave the performance shown in Table 8.2.

Table 8.2. PERFORMANCE OF 'CONSTANT SENSITIVITY' BRIDGE

Temperature (°C)	14·8	15·2	18·0	24·0	27·2	28·2
$\frac{dI_g}{dT}$ (μA K^{-1})	2·45	2·45	2·47	2·46	2·45	2·45

In this work theoretical and practical results differ by about 1%, which is tolerable for most purposes.

8.2.1.4. SELF-BALANCING THERMOMETER

Priestley[11] has described a self-balancing bridge in which the d.c. out-of-balance signal from the bridge is fed to a servo-amplifier, which drives a servo-motor, which is connected to a balancing ten-turn potentiometer in one arm. This has a straight-through spindle and one end of this is connected to a Digidial. Priestley incorporated the thermistor in shunt with a resistance S of value

$$S = R_0 \frac{(B-2T_0)}{(B+2T_0)} \tag{8.14}$$

which gives a second differential coefficient of zero in the Taylor expansion of the resistance of the combination as a function of T about T_0. R_0 is the thermistor resistance at T_0 and B the material constant. (cf. Beakley's[7] corresponding criteria for a resistance in series with the thermistor).

An earlier thermometer for measuring small temperature changes on a short-term basis was described by Greenhill and Whitehead,[5] who achieved an accuracy of 0·0002 °C in 0·01 °C. The sensing thermistor was incorporated in a d.c. Wheatstone bridge with a reflecting galvanometer detector. The incident light to this was chopped using a slotted rotating disc to produce a.c. at about 5 kHz. The reflected light impinged on a photoelectric cell after passing the edge of a wide slit, was amplified and produced an output which was fed back to a second thermistor in an adjacent arm, thereby heating it. The resistance of this second thermistor depends on the amplitude of the a.c. output from the amplifier, which in turn depends upon the amount of off-balance current in the galvanometer. The output a.c. of the amplifier is recorded an after calibration of the sensing thermistor can be used as a measure of temperature change.

8.2.2. Differential Thermometers

For some applications the measurement of temperature difference, rather than a direct measurement of temperature, is required. A Wheatstone bridge network is again to be preferred, but now incorporating two thermistors in adjacent arms as sensing elements. Ideally, these should be perfectly matched, not only at the operating temperature, but over the range of temperature difference to be measured; this condition does not arise naturally so equalisation is necessary.

In their bridge Nordon and Bainbridge[12] first linearised each thermistor using shunt resistors whose values were given by Equation 8.13, and then equalised the combined resistance at a given temperature, by connecting a resistance in series with one of the combinations. The different temperature sensitivities were compensated by passing different constant currents through each arm, of such values that the same potential variations per unit of temperature change occurred across each. A high impedance detector was used in the bridge to permit this. A linear dependence on temperature, with a maximum error of 0·5% was obtained over a range of 50 K and with a maximum error of 0·05% over a range of 18 K about ambient.

Godin[13] used a somewhat different approach to equalisation. Only one of the thermistors was shunted by a resistance and the value of this was determined by equating the first temperature-coefficient of resistance of the combination with that of the other thermistor. Equalisation of resistance at the operating temperature was achieved by connecting a small resistance either in series with the shunt combination or the other thermistor. Gordon and Seay[14] were concerned with the equalisation of a group of thermistors having a 5% tolerance at a given temperature. Their solution, which is also applicable to a pair of thermistors, was to employ a parallel resistance R_a, such that the combined resistance is $R_t = RR_a/(R+R_a)$, so that

$$\frac{dR_t}{dR_a} = \frac{R_a^2}{(R_a+R)^2} \tag{8.15}$$

and hence

$$R_a = \frac{R}{\left\{-1+\left(\dfrac{dR}{dR_t}\right)^{1/2}\right\}} \tag{8.16}$$

which is used to obtain a required slope of the R_t/T characteristic. Again, series resistors are added to produce spot-temperature alignment.

Godin[15, 16] described the design of two differential bridges. The first used two uncompensated thermistors within the bridge and was based on analysis for zero temperature-coefficient of detector galvanometer current and maximum sensitivity to differential temperature changes. This involved non-linear scaling and a variation of sensitivity with temperature, unless operation was close to the design temperature. In the second design equalisation was considered.

Further representative, but not exhaustive, references to the measurement of temperature, including that of air, soil, sea and the human body are given at the end of this chapter[17–25].

8.2.3. Temperature Stability

It is of paramount importance in a measuring instrument that the scale reading does not change significantly after calibration. Thermistors used for thermometry in particular should therefore be aged[1, 26]. The following is a summary of the conclusions which may be drawn concerning long and short term accuracy, which has been culled from the literature[1].

8.2.3.1. N.T.C. THERMISTORS

(a) Measurement of small temperature differences over short periods of time can be very accurate. In a calorimeter 0·0002 K has been claimed[26] in a temperature difference of 0·01 °C. A short-term stability of 5 parts in 10^6, corresponding to an accuracy of temperature of 0·0001 K has also been reported.

(b) For larger temperature differences, over a short period of time, Beck has quoted an accuracy of the order of 0·01 °C in a temperature difference of 6–12 °C.

(c) Long-term stability depends on the previous history of the thermistor and whether it has been 'aged'. It may be affected by thermal shock or large temperature changes.

(d) A short-term stability of ±0·2 K has also been achieved for thermistors with especially-low resistivity[27], made to operate at liquid-nitrogen boiling point (−253 K). However, if long-term stability better than ±10 °C is required periodic recalibration is necessary.

11

8.2.3.2. P.T.C. THERMISTORS

The only information at present available for perovskite types is due to Saburi and Wakino (see Section 2.3.1). Although detailed information on Si thermistors is not available, it is to be expected that stability should be very good.

8.3. TEMPERATURE SENSING AND CONTROL[1, 27–38]

There are many applications in which the temperature dependence of thermistor resistance is used as an indicator of temperature, rather than as a means of measurement. The principle of operation does not differ, only the way in which the output signal is handled. In control applications the error signal is generally amplified and either operates a relay (ON-OFF control) or controls the conduction period of a thyratron, silicon-controlled-rectifier or the operation of a power transistor.

The large temperature coefficient of resistance of barium titanate type thermistors permits their use in many cases as direct control elements, without amplification. A simple example is shown in Fig. 8.3: here a 'Positor' shunts a neon lamp, whose ignition indicates temperature in a fire-alarm system. Another is shown in Fig. 8.4: a 'Positor' is connected in series with a heater to control the temperature of a confined space. For direct control in this way the power consumed in the thermistor must be limited so that self-heating above T_p (see Fig. 2.13) does not occur.

Because of the wide range of operating temperatures for which P.T.C. thermistors are available (approximately - 100 °C to +350 °C) they find widespread use in sensing both rising and falling temperatures. The positive temperature coefficient of resistance makes a simple fail-safe relay circuit possible, that can switch off a temperature-sensitive load, not only in the event of overheating, but also if the current supply to the protective circuit should be interrupted.

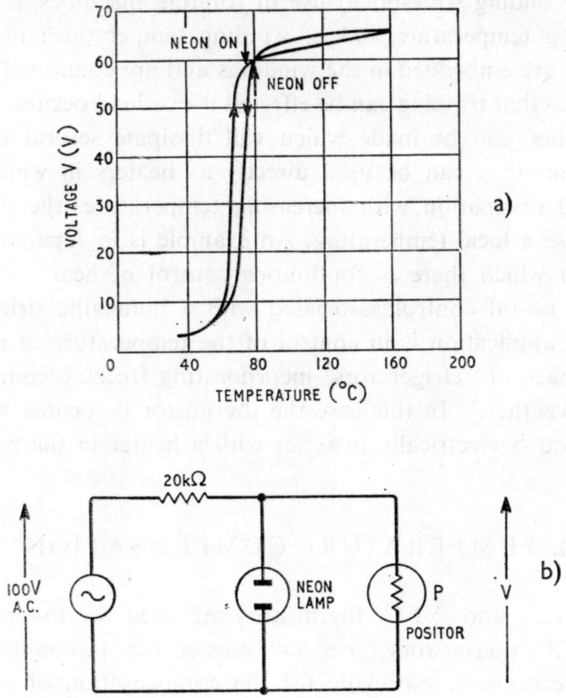

Fig. 8.3. Application of a Positor to a fire-alarm system (after Saburi and Wakino)

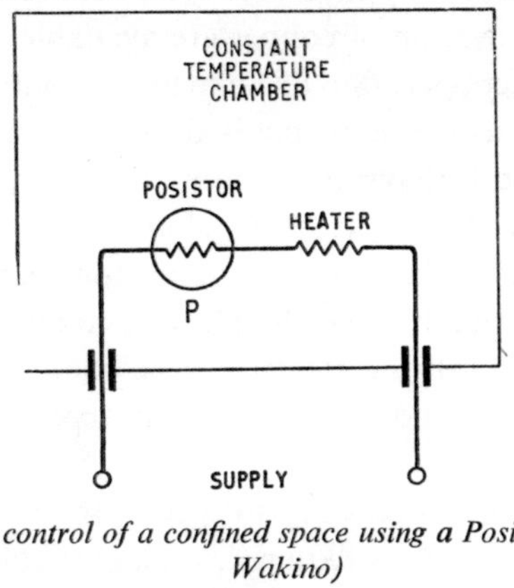

Fig. 8.4. Temperature control of a confined space using a Posistor (after Saburi and Wakino)

They are finding widespread use in rotating machines as indicators of bearing temperature and of winding temperature: in the latter case they are embedded in the windings and are connected in control circuits, so that tripping can be effected if overload occurs.

'Positors' can be made which will dissipate several watts. This means that they can be used directly as heaters in which there is decreased dissipation with increasing temperature, thereby tending to stabilise a local temperature. An example is in Japanese footwarmers[37] in which there is continuous control of heat output rather than the on-off control associated with a bimetallic strip. Another Japanese application is in control of the temperature of the refrigeration space of refrigerators, incorporating freezing compartments, in cold weather.[37] In this case the thermistor is located near to the freezer and is electrically in series with a heater in the refrigeration space.

8.4. TEMPERATURE COMPENSATION[1, 39–46]

Both N.T.C. and P.T.C. thermistors are used for this purpose. To date N.T.C. thermistors have had greater use, in conjunction with passive resistances, especially for the compensation of the positive temperature coefficient of resistance of copper ($+0{\cdot}39\%$ per °C at 25 °C): examples are the resistances of wire-wound meter movements[39–42] and cathode ray tube focus and deflection coils. Tabulated data for the compensation of copper are available. Below a graphical design method for copper, due to Bennett,[40] is outlined.

The design is based on a normalised set of resistance/temperature curves, for thermistors having a common temperature coefficient of resistance of $-3{\cdot}86\%\ °C^{-1}$ at 25 °C, shunted by a resistance of 1 Ω. This is shown in Fig. 8.5 for five thermistors having resistances at 25 °C of 0·4, 0·6, 1·0, 1·4 and 2·4 Ω. The compensating network is connected in series with the element to be compensated. Subsequent scaling of both shunt and thermistor resistances is required to meet a specific design.

For any problem it is necessary to specify the range of temperature over which compensation is required. As a particular example let this

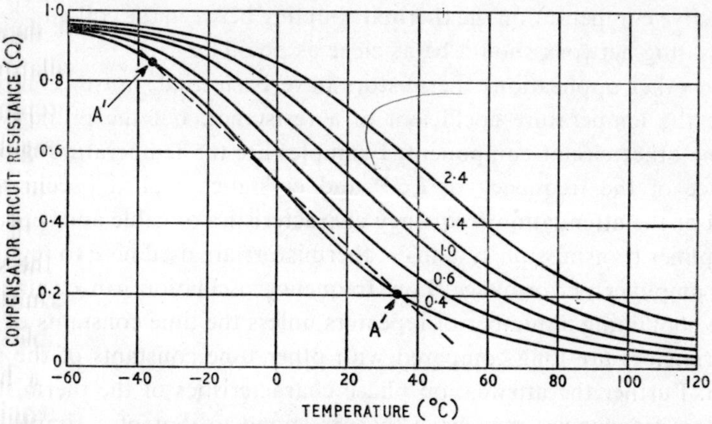

Fig. 8.5. Temperature dependence of the terminal resistance of a parallel combination of a 1 Ω resistance and an N.T.C. thermistor, having a temperature coefficient of resistance of −3·86% °C⁻¹; thermistor resistance (Ω) at 25 °C is the parameter (after Bennett)

be −40 to +40 °C. The first step in the design is to construct a straight line, superposed on one of the characteristics of Fig. 8.5, that comes closest to fitting it. In making a choice the aim should be that the maximum deviations at the limits and at two equally spaced intermediate temperatures should be the same. In this case the best fit is obtained for the 0·4 Ω thermistor, as shown by the line AA'. It should be noted that the change of compensator-circuit resistance between A and A' is 0·755 Ω.

If the actual coil to be compensated has a resistance of 100 Ω at 25 °C, then its overall change of resistance between the extreme temperatures of −40 and +40 °C is approximately given by

$$\Delta R = 100 \times 80 \times 0{\cdot}0039 = 31{\cdot}2\ \Omega$$

which is also the magnitude of the required change in resistance of the compensating circuit. Hence the compensating-circuit values must be scaled by a factor 31·2/0·755 = 41·4. It follows that the actual value of shunt resistance required is 41·4 Ω and the appropriate thermistor resistance at 25 °C should be 0·4×41·4 = 16·56 Ω. For effective compensation the thermal coupling between the coil and compensating network should be as close as possible.

In other applications thermistors have been used, not to compensate the temperature coefficient of a resistance, but the property of some other circuit component. Examples are the temperature dependence of the frequency of *LC*[43] and crystal-controlled oscillators[44] and of the attenuation/frequency characteristics of cable and repeater amplifier transmission systems[45]: thermistors are used here to regulate the amplifier performance. Low frequency oscillation can occur in a line containing a number of repeaters unless the time constants of the thermistors are long compared with other time constants of the system. Further the attenuation/phase characteristics of the thermistors at low frequencies may have to correspond to that of a simple *RC* network. Special N.T.C. directly-heated thermistors with a high ratio of time constant to mass have been developed to give the required stability and sensitivity.

Further compensation applications have been in transistor bias circuits[46] and photographic exposure meters incorporating selenium cells.

REFERENCES

1. SCARR, R. W. A. and SETTERINGTON, R. A. 'Thermistors, their theory manufacture and application', *Proc. Instn. elect. Engrs.*, **107B**, 395 (1960)
2. ANDREWS, B. L. 'The Thermistor in biological research', *Electron. Engng.*, **19**, 288 (1947)
3. DEETER, E. L. 'Null temperature bridge', *Electronics*, **21**, 180 (1948)
4. HERINGTON, C. F. G. and HANDLEY, R. 'The use of thermistors for recording small temperature differences', *J. Scient. Instrum.*, **25**, 434 (1948)
5. GREENHILL, E. B. and WHITEHEAD, J. R. 'An apparatus for measuring small temperature changes in liquids' *J. Scient. Instrum.* **26**, 92 (1949)
6. MACRADYEN, A. 'A simple device for recording mean temperatures in confined spaces' *Nature*, **164**, 965 (1949)
7. BEAKLEY, W. R. 'The design of thermistor thermometers with linear calibration'. *J. Scient. Instrum.*, **28**, 176 (1951)
8. MCLEAN, J. A. 'A method for constructing direct-reading thermistor thermometers' *J. Scient. Instrum.*, **31**, 455 (1954)
9. COLE, K. S. 'Thermistor thermometer bridge: linearity and sensitivity for a range of temperature'. *Rev. Scient. Instrum.*, **28**, 326 (1957)
10. PITTS, E. and PRIESTLEY, P. T. 'Constant sensitivity bridge for thermistor thermometers', *J. Scient. Instrum.*, **39**, 75 (1962)

11. PRIESTLEY, P. T. 'A thermistor thermometer with digital display', *J. Scient. Instrum.*, **40,** 505 (1963)
12. NORDON, P. and BAINBRIDGE, N. W. 'The use of unmatched thermistors for the measurement of temperature difference under varying ambient conditions', *J. Scient. Instrum.*, **39,** 399 (1962)
13. GODIN, M. C. 'A method of equalising thermistors', *J. Scient. Instrum.* **39,** 241 (1962)
14. GORDON, W. E. and SEAY, P. A. 'Compensation of thermistor elements', *Bull. Am. met. Soc.*, **29,** 41 (1948)
15. GODIN, M. C. 'Simple bridge for the direct measurement of temperature differences', *J. Scient. Instrum.*, **38,** 330 (1961)
16. GODIN, M. C. 'A modified thermistor bridge', *J. Scient. Instrum.*, **40,** 500 (1963)
17. COUTTS, J. R. H. 'Observation of soil temperature in North East Scotland', *Sixième Congrès de la Science du Sol*, Paris (1956)
18. BADGLEY, F. I. 'Response of radiosonde thermistors'. *Rev. Scient. Instrum.* **28,** 1079 (1957)
19. LEISS, W. J. and TRUFANT, R. F. 'Improved bathythermograph and the measurement of ocean temperature', *Sound*, **2,** 42 (1963)
20. LA FOND, E. C. 'Towed sea temperature structure profiler', *Mar. Sci. Instrum.* **2,** 49 (1963)
21. HALES, W. B. 'Thermistors as instruments in thermometry and anemometry,' *Bull. Am. met. Soc.*, **29,** 494 (1948)
22. HAUCK, V. D. COSBY, J. R. and DEMBER, A. B. 'Radiosonde telemetering systems', *Electronics*, **19,** 120 (1946)
23. KROG, J. 'Thermistor hypodermic needle for subcutanious temperature measurement', *Rev. Scient. Instrum.*, **27,** 408 (1956)
24. TSIEN, H. C. 'Piston zone temperature measurement', *Instrums Control, Syst.*, **37,** 105 (1964)
25. FINCH, A., GARDNER, P., LANE, R. D. G., and SMETHURST, P. 'A precision cryometer', *Lab. Pract.*, **14,** 448 (1965)
26. BECK, A. 'The stability of thermistors', *J. Scient. Instrum.*, **33,** 16 (1956)
27. SACHSE, H. B. 'Thermistors as sensing elements for low temperatures', *Bull. Refrig. Inst.*, **1,** 145 (1958)
28. BATES, S. 'Bearing temperature control', *Tech. J. Brush Grp*, **2,** 2 (1954)
29. HOWES, T. W. 'Characteristics and applications of thermally-sensitive resistors or thermistors', *Electl. Commun.*, **32,** 98 (1955)
30. DICKINSON, C. J. 'Thermistor continuous temperature control for biological research', *Electron. Engng.*, **21,** 408 (1949)
31. TAYLOR, A. H. 'Thermistor thermostats', *Electronics*, **23,** 154 (1950)
32. VODDEN, H. A. 'A thermostat control unit', *J. Soc. Chem. Ind, Lond.*, **69,** 51 (1950)
33. TEMPEST, W. 'A bridge thermostat with a differential of 0·001 °C, *Electron. Engng.*, **35,** 814 (1963)
34. HAY, J. N. 'A thermostat control unit for a polarizing microscope', *J. Scient. Instrum.* **41,** 465 (1964)
35. MANYAKIN, G. I., and PETROV, V. I. 'Semiconductor temperature controller', *Instrum. Constr.*, **11,** 31 (1964)

36. THORNE, S. W. 'Temperature control system for use with a spectrophotometer', *J. Scient. Instrum.*, **37**, 478 (1960)
37. IMAMURA, E., and TAMADA, M. 'Latest applications of Murata Posistor', *Japan Electron. Engng.*, **12**, 60 (1967)
38. ANDRICH, E. 'Properties and applications of P.T.C. thermistors', *Electron. Applic.*, **26**, 123 (1966)
39. GENERAL ELEC. CO. METALLURGICAL PRODUCT DEPARTMENT. 'A thermistor selecting shortcut', *Control Engng.*, **4**, 73 (1957)
40. BENNETT, F. 'Designing thermistor correcting networks graphically', *Control Engng.*, **2**, 66 (1955)
41. PATTEE, T. R. 'Temperature compensation of instruments with negative coefficient resistors', *Electronics*, **16**, 102 (1943)
42. ILYNKOVICH, A. M. 'Compensation of the temperature error of induction meters and wattmeters by thermistors', *Elektrichestvo*, **8**, 76 (1960)
43. GUENÉ, P. 'The use of thermistors for the compensation of drift in self-oscillating circuits', *Anns Radio*, **11**, 317 (1956)
44. ANON., 'Temperature-compensated crystal oscillator' *Electron. Radio Engr.* **35**, 314 (1958); also *Wireless World*, **64**, 441 (1958)
45. BOLLMAN, J. H. 'A pilot channel regulator for the K1 carrier system', *Bell Labs. Rec.* **20**, 258 (1942)
46. WHEELER, A. J. 'Thermistors compensate transistor amplifiers', *Electronics*, **30**, 169 (1957)

CHAPTER 9

Miscellaneous Applications of Directly-heated Thermistors

9.1. INTRODUCTION

In this chapter various applications of both N.T.C. and P.T.C. thermistors will be considered, which are based on either their non-linear or transient properties. Non-linear applications include oscillator amplitude and/or frequency regulation, amplifier gain or level stabilisation, voltage and current limiting or regulation, volume limiting and signal expansion and compression. Transient applications are in overload protection, surge suppression, time delay in relay circuits and switching.

9.2. OSCILLATOR AMPLITUDE AND/OR FREQUENCY STABILISATION

The essentials of a conventional feedback oscillator circuit are a frequency-defining network and a high-gain electronic amplifier. In an ideal circuit oscillation should occur at a stable amplitude at a frequency solely determined by the resonant structure, and the oscillation should be independent of changes within the amplifier. The 'Meacham bridge' circuit[1] shown in Fig. 9.1 provides both frequency and amplitude stabilisation and since it operates with the amplifier in a linear regime it delivers a very pure sinusoidal output.

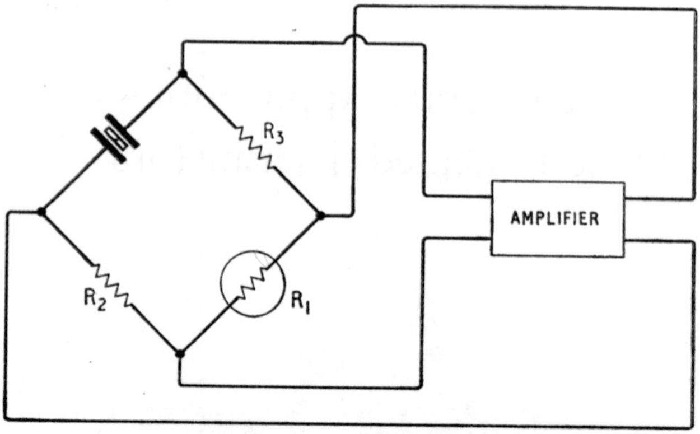

Fig. 9.1. The Meacham-bridge circuit

The bridge is comprised of the frequency-defining element—a series-resonant crystal in Meacham's original circuit—of impedance $Z_4 = R_4+jX_4$, two passive resistances R_2 and R_3 and a thermally-sensitive resistance R_1. The amplifier, which feeds signal to the bridge and in turn receives unbalance signal *e* from the bridge at its input, provides phase-reversal between input and output. For exact reactive balance of the bridge $X_4 = 0$, and for exact resistive balance $R_1 = R_2R_3/R_4$. For oscillation to occur R_1 must have a slightly smaller value than this so that the bridge is slightly unbalanced. The attenuation of the bridge just compensates the gain of the amplifier; in other words the loop gain is unity. The use of fixed resistors for R_1 as well as R_2 and R_3 would cause a circuit performance critically-dependent on component tolerances: the circuit might either fail to oscillate or build up such an amplitude that the amplifier overloads. By choosing R_1 to have a positive temperature coefficient of resistance, however, a working solution is obtained. On initially switching on, the cold resistance of R_1 is well below the value required for balance, so that the bridge attenuation is relatively small and oscillation builds up rapidly.

As the thermistor heats up, due to the power dissipation within it, so its resistance increases and its value approaches that required to give a loop gain of unity, and amplitude stability is achieved. If for some reason R_1 becomes too big, the unbalance potential either becomes too small to support the level of the oscillation or even reverses in phase, so that the amplitude automatically decreases.

Frequency stability approaches that of the frequency-defining-network if the amplifier gain is large. If there is a change in the phase-response of the amplifier this must be matched by an equal and opposite phase change in the output voltage e from the bridge. Since the magnitudes of the voltages across each of the bridge arms greatly exceed $|e|$, the required phase-change in e can be obtained for a relatively small change in $|Z|$, and hence in departure from series resonance of the crystal.

A feature of this design of oscillator is that the amplifier itself acts in a linear regime, since it does not have to provide limiting action by its own non-linearity, as is the case with oscillators not employing an external non-linear element for this purpose. The frequency of operation has to be considerably greater than the reciprocal of the effective circuit time constant of the thermistor, whose resistance must not vary during the cycle of oscillation. In consequence it does not generate harmonics, whose intermodulation would result in fundamental frequency instability.

In Meacham's original work R_1 was provided by a small tungsten filament lamp of low wattage—such lamps have a temperature coefficient of resistance of $\approx +0{\cdot}5\%$ per K. The development of P.T.C. bulk thermistors now permits their use in the basic circuit of Fig. 9.1. By a simple change of circuit form N.T.C. thermistors can also be used: either R_2 or R_3 is made the N.T.C. element.

Meacham's work was confined to fixed frequency oscillators. Shepherd and Wise[2] extended this to variable-frequency oscillators. They considered two types, based on the twin-T RC and bridged-T LC 'null transmission' networks, which replaced the Wheatstone-bridge network used by Meacham, as shown in Fig. 9.2a and b, respectively. In the twin-T case oscillation occurs at approximately $f_0 = 1/2\pi R\sqrt{(2C_1C)}$ and it is necessary that R_1 be less than $4C/RC_1$ for positive feedback and oscillation to occur—see also Hyde[3]. Minimum rather than null transmission then occurs and the phase-shift through the filter is approximately 180°, giving the necessary positive feedback required for an antiphase-transmission amplifier. The value of R_1 determines the attenuation through the twin-*T* filter at the minimum and this should again be such that the product of the attenuation and

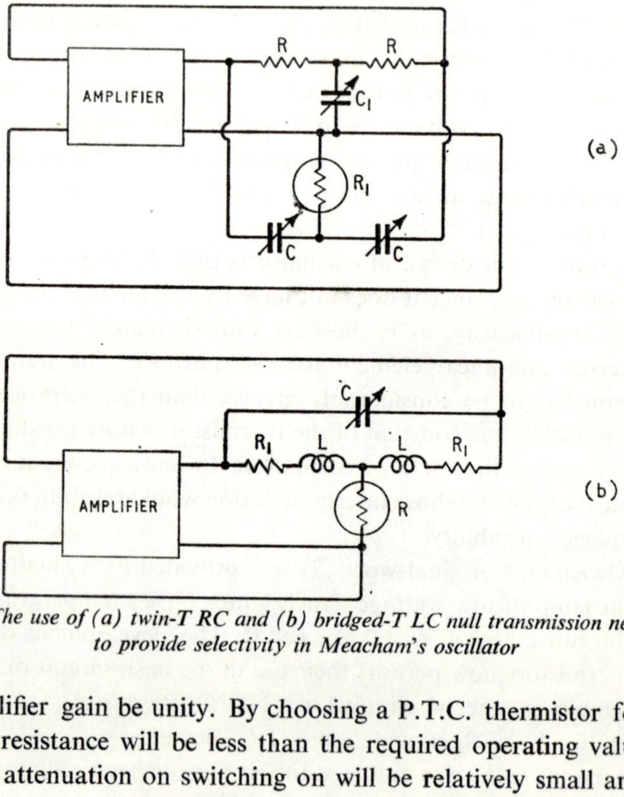

Fig. 9.2. The use of (a) twin-T RC and (b) bridged-T LC null transmission networks to provide selectivity in Meacham's oscillator

the amplifier gain be unity. By choosing a P.T.C. thermistor for R_1, its cold resistance will be less than the required operating value, so that the attenuation on switching on will be relatively small and the amplitude of the oscillation builds up rapidly. Heating of the thermistor by internal power dissipation brings its resistance to the required operating value. The operation of the *LC* circuit of Fig. 9.2b is similar. Other variable frequency oscillators have been described, incorporating as the frequency-defining network an *RC* phase-shift circuit a Wien-bridge and an *LC* network.[4–6]

9.3. AMPLIFIER AUTOMATIC GAIN CONTROL

The incorporation of either an N.T.C. or P.T.C. thermistor in the negative-feedback loop of an amplifier provides a means of stabilising the gain[7-9]. A schematic illustration of the basic circuit arrangement for an N.T.C. thermistor is shown in Fig. 9.3. As the amplifier output falls so the resistance of the thermistor rises and the amount

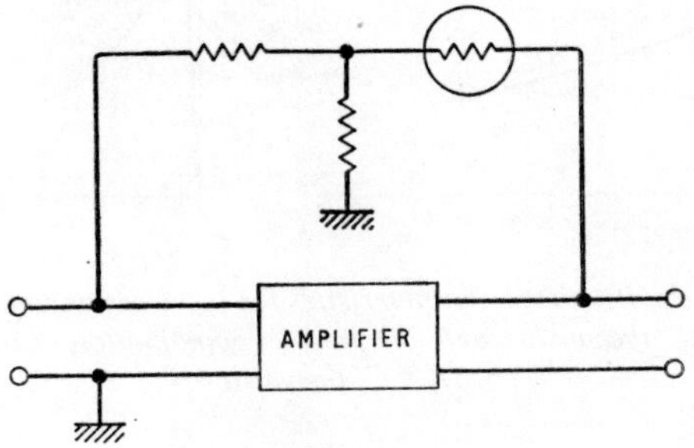

Fig. 9.3. Schematic illustration of amplifier gain control using an N.T.C. thermistor in a feedback loop

of signal feed-back is reduced, thereby compensating the original change. A P.T.C. thermistor could be used as the shunt element in the feedback path. For some applications of this kind the indirectly-heated thermistor has some advantages (see Section 10.4.4)

9.4. REGULATORS OR LIMITERS, AND RELATED APPLICATIONS[10-13]

Typical current-voltage characteristics for a series combination of passive resistor and N.T.C. thermistor and a parallel combination of passive resistor and P.T.C. thermistor are shown in Fig. 9.4. In each case a single thermistor is involved, with ambient temperature and dissipation constant invariant. The families of characteristics are generated by choosing different values of associated passive

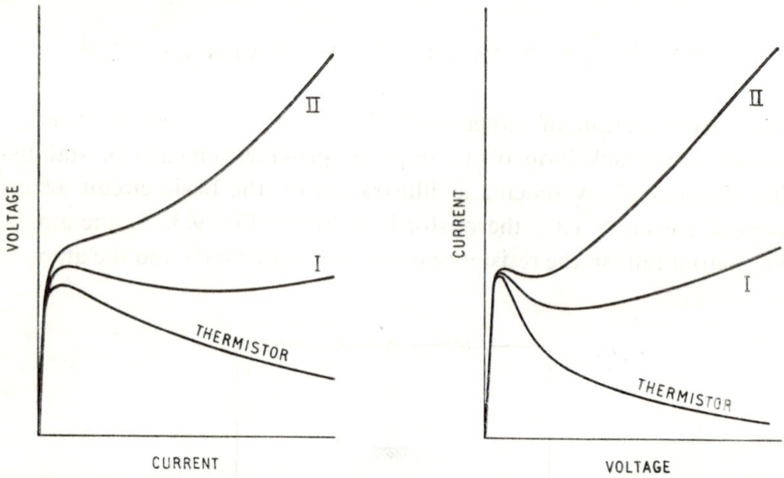

Fig. 9.4. Typical current/voltage characteristics for (a) a series combination of passive resistor and N.T.C. thermistor and (b) parallel combination of passive resistor and P.T.C. thermistor

resistances. For resistances approximately equal to the mean slope resistance of the thermistor characteristic beyond turnover, the combined characteristics (curves I) are substantially independent of voltage and current, respectively, so that the two combinations can be used for voltage and current regulation (or limiting).

N.T.C. thermistors found early application for volume-limiting of speech signals. Combined characteristics, II, can similarly be used to 'compress' a range of signals rather than limit it. In both limiting and compression the N.T.C. thermistor and series resistor are placed in parallel with the load and the P.T.C. thermistor and shunt resistor in series with it.

If the N.T.C. thermistor and series resistance are connected in series with the load or the P.T.C. thermistor and shunt resistance in parallel with the load then correspondingly an 'expandor' is created.

In the above statements it has been assumed that the signals being controlled have frequencies sufficiently low that there is no significant departure from an operating point on the combined static characteristics. For changes which are rapid in relation to the effective circuit time constant the operating point moves off the static characteristic along an ohmic resistance line from the origin. For such fluctuations control action may then be derived from the resistance changes resulting from the mean square power dissipated in the thermistor. The selective action of the circuits on a frequency basis can be used to advantage in speech volume limiters. Here a choice of thermistor is made so that limiting action is produced for changes in volume which are syllabic in frequency or slower, but the thermistor

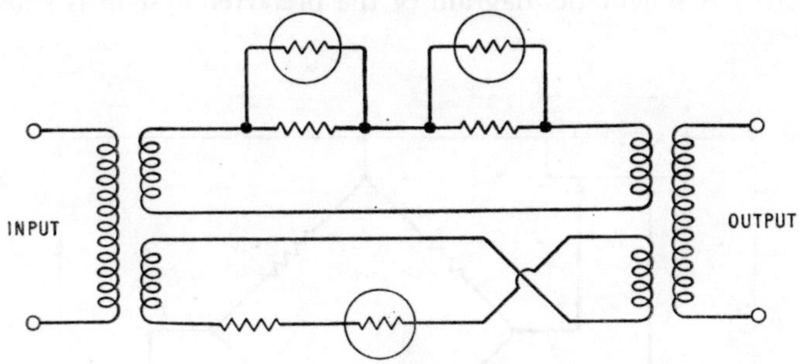

Fig. 9.5. Speech-volume limiter circuit (after Weller)

does not follow the more rapid speech variations, so that wave-form distortion does not arise. Speech volume limiters can accommodate large volume changes without introducing wave-form distortion[10, 11]. That due to Weller[13] is shown schematically in Fig. 9.5. It is a hybrid coil arrangement with the input connected to a transformer with balanced output windings. The signal is divided into two paths and passes to a similar transformer on the output side. The secondary windings of the output transformer are of reversed magnetic polarity so that the primary currents of the two paths introduce opposing currents in the output line. If both paths had equal resistance there would be no output signal. However, at low signal levels the N.T.C. thermistor resistance is very high and that of the P.T.C. element is low. In this condition most of the signal current passes through the low resistance path, the bridge is unbalanced and the circuit introduces little loss in the line. At signal levels in the limiting range the resistances in the two paths converge, so introducing more loss in the bridge.

9.5. PRECISION A.C. VOLTAGE STABILISERS

The use of thermistors in precision a.c. voltage stabilisers with output voltages constant to the order of 0·01% has been described by Patchett[14]. A schematic diagram of the preferred system is shown in

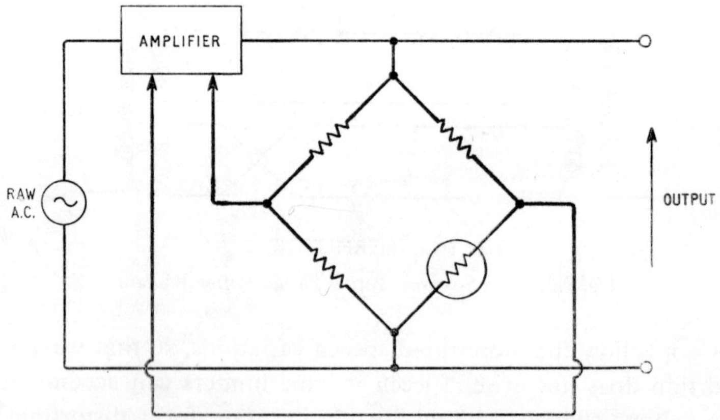

Fig. 9.6. Schematic circuit of a.c. voltage stabiliser (after Patchett)

Fig. 9.6. The thermistor forms one arm of a Wheatstone bridge connected to the stabilised supply, the other arms being passive resistors. The bridge is balanced for the prescribed output voltage and any change in output causes unbalance of the bridge; the error signal is fed to the phase-sensitive controller (amplifier), which acts to correct the signal change. Patchett[15] has shown that the figure of merit, F, for an indicator bridge incorporating a non-linear element in one arm, defined as the change in output voltage to the change in supply voltage, is

$$F = \frac{P(n-1)}{(np-1)(p+1)} \tag{9.2}$$

where $p = R_1/R_2$ is the bridge ratio and n is defined by

$$V = kI^n \tag{9.3}$$

where V and I are the terminal voltage and current, respectively, for the non-linear element and k is a constant. In the negative resistance region the N.T.C. thermistor has a characteristic approximating closely to Equation 9.3. From Equation 9.2 it follows that F is infinite if $np = 1$; this condition can be satisfied with an N.T.C. thermistor if $p \approx 2$. In physical terms this situation corresponds to operation of the thermistor and its equivalent series resistance presented by the bridge at the minimum of the composite characteristic shown for limiter action in Fig. 9.4.

For this particular application it is necessary that the thermal time constant of the thermistor be chosen so that it can follow the fluctuations of the system. It must not be so short, however, that there is significant variation of its resistance over the period of the voltage which is being controlled. The limitations on performance can be calculated from the analysis of large-signal behaviour of thermistors given by Ekelöf et al[16] (see Section 4.3).

Because the resistance of the thermistor depends on ambient temperature, bridge unbalance can arise because of fluctuations in this. Patchett has drawn attention to the significance of such fluctuations in creating a false error voltage, incorporating harmonics of the fundamental frequency being controlled. He has also shown[15] how the dependence of the resistance of the bridge thermistor on ambient temperature can be corrected by incorporating another N.T.C. thermistor of large thermal capacity in the feed-circuit to the bridge.

9.6. DELAY AND SURGE SUPPRESSION

The thermal inertia of both N.T.C. and P.T.C. thermistors is made use of in delaying the application of a signal to other circuit elements or in protecting them from transient surges.

Examples are the protection of thermionic lamps or of valve heaters[11, 17, 18] and of fluorescent lamps.[11] Other applications include the starting of electric motors[12, 19] and the suppression of telephone bell tinkle[20]. The connection of a single P.T.C. thermistor in series with a series string of lamps will give protection against a short-circuit failure in any one lamp and similarly against short-circuit failure when connected in series with the capacitor controlling a fluorescent lamp. A series connection of lamps can also be protected against open-circuit failure if an N.T.C. thermistor is connected in

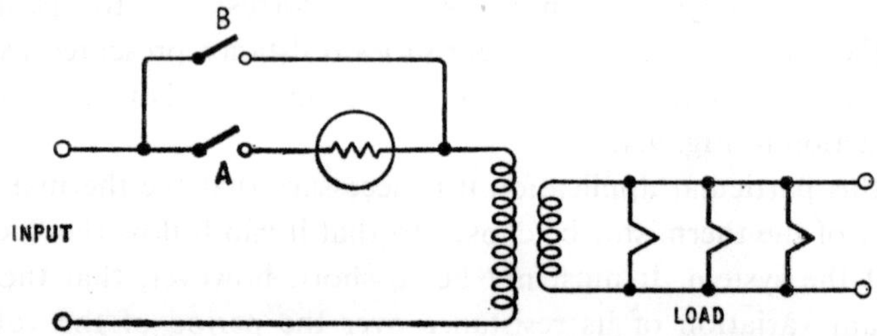

Fig. 9.7. Use of thermistors to control valve heater powers

parallel with each one. The thermistors normally pass little current: if a lamp goes open-circuit however, then virtually the whole line voltage will be thrown instantaneously across its thermistor, whose resistance will fall until it carries approximately the normal line current and the remaining lamps stay alight.

Gano and Sandy[17] have described the use of N.T.C. thermistors to control valve heater powers in excess of a kilowatt. The basic circuit is shown in Fig. 9.7. The current ratings of the thermistors make them suitable for inclusion in the primary circuit of the transformer. The voltage is initially applied to the filaments by closing contact *A*. The thermal response of the thermistors is chosen to be much slower than that of the heaters so that the relatively slow drop in thermistor resistance causes the current in the filaments to rise gradually. After an equilibrium voltage is reached the thermistors are shorted out by closing contact *B*. From a cold start to the time of shorting the resistance of the thermistors decreases by a factor of about 100 and the heater resistance increases by a factor of 5 to 10. Time delay circuits for relays can be made with delays from a few seconds to several minutes. N.T.C. thermistors are connected in series and P.T.C. thermistors in parallel with the actuating winding for this purpose. In view of the sensitivity of thermistor resistance to ambient temperature it is evident that they are particularly suited where the achievement of a minimum delay is required, rather than a precisely defined one. Examples are the switching on of valve anode voltage after the cathode has reached operating temperature, in the interval circuit for record

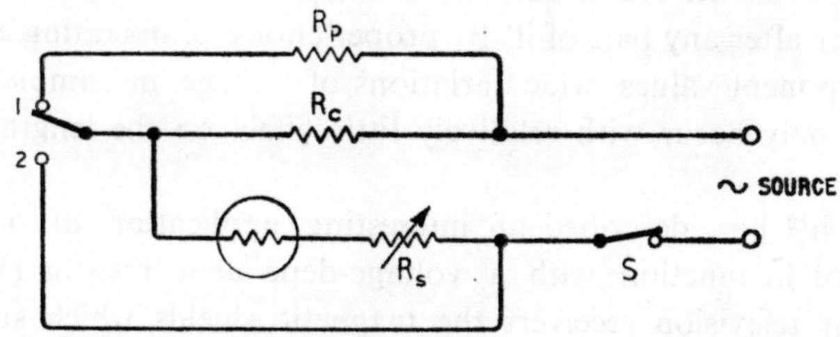

Fig. 9.8. Circuit for delay in operation of an a.c. relay, which incorporates an anti-chatter facility

changers and the switching on and off of advertisement illumination. For repetitive operation it is necessary to effect electrical disconnection of the thermistors from the circuit after each operation. This is accomplished by incorporating an extra working contact on the relay.

In Fig. 9.8 a circuit is shown for producing a delay in the operation of an a.c. relay which incorporates an anti-chatter facility[21]. After the switch *S* has been closed, with the relay working contact in position 1, current flows through the series resistance R_s and the N.T.C. thermistor and then through the parallel combination of the relay coil and R_p. R_p is a factor of 3 to 10 smaller than the resistance R_c of the coil. As the resistance of the thermistor falls the current gradually increases until the armature of the relay undergoes the first vibratory movement: this has to be sufficient to switch the working contact from 1 to 2. The instant this contact switches over the total current in the circuit drops, but the current in the relay coil rises so giving rise to a sure movement of the relay. To avoid the limitations imposed by ambient temperature in designing for a wide timing interval and for a lengthy cooling time between operations of an N.T.C. thermistor operated initially cold, a mode of operation has been used in which two relays or the equivalent are used and the thermistor is initially heated to a low-resistance value by passing a relatively large current through it for an interval which is short compared with the desired time delay[10]. The current then reduces automatically as the thermistor cools and eventually the working relay is tripped. The thermistor is available for re-use immediately after a completed timing interval or after any part of it. By proper choice of operating currents and component values, wide variations of voltage and ambient temperature may occur with relatively little effect on the length of the delay time.

Andrich[12] has described an interesting application of a P.T.C. thermistor in junction with a voltage-dependent resistor (V.D.R.) In colour television receivers the magnetic shields which surround shadow-mask tubes to prevent beam-landing errors due to extraneous magnetic fields are provided with degaussing coils. Ideally these should be energised each time the set is turned on and then gradually de-energised. The circuit is shown in Fig. 9.9. When the set is switched on the thermistor resistance is low so that a large current is passed through the coils and a magnetomotive force of some 350 At is developed. As the thermistor heats up its resistance rises and the magnetomotive force is reduced to 0·03 At. When the mains voltage is first

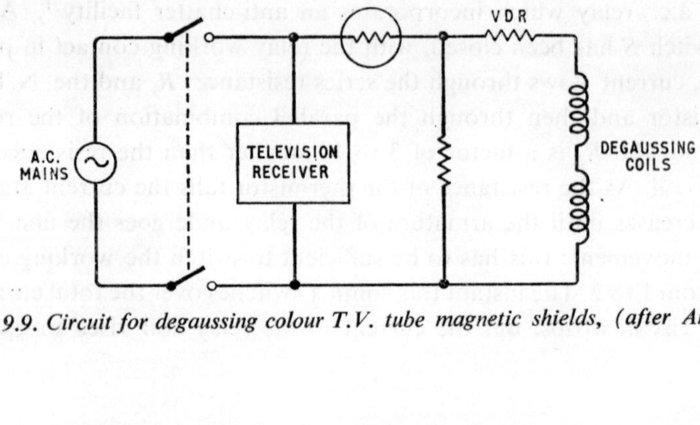

Fig. 9.9. Circuit for degaussing colour T.V. tube magnetic shields, (after Andrich)

applied the V.D.R. has a low resistance so that about 90% of the current flows through the degaussing coils. As the voltage across the V.D.R. falls so its resistance rises and the end result is a small steady-state current flowing through the shunt resistor.

There is an upper limit to the magnitude of a surge which can be suppressed by a single thermistor, which is governed by its mass: for a particular mass, beyond a given input power there is a tendency for the thermistor to shatter due to transient thermal stresses[22]. The maximum instantaneous power dissipation which can be accommodated is of the order of 500 W.

9.7. INTERCONNECTION OF N.T.C. AND P.T.C. THERMISTORS[12, 23, 24]

In Section 5.3.3 it was pointed out that the combination of negative resistance, inductance and capacitance associated with an interconnection of an N.T.C. and a P.T.C. thermistor can result in oscillation. This is essentially of the relaxation variety. A simple circuit and the intersection of the thermistor characteristics to define a suitable operating point are shown in Fig. 9.10; the load resistance is ignored. Not all cases in which the curves intersect in the negative resistance will lead to oscillation because the available negative resistance is

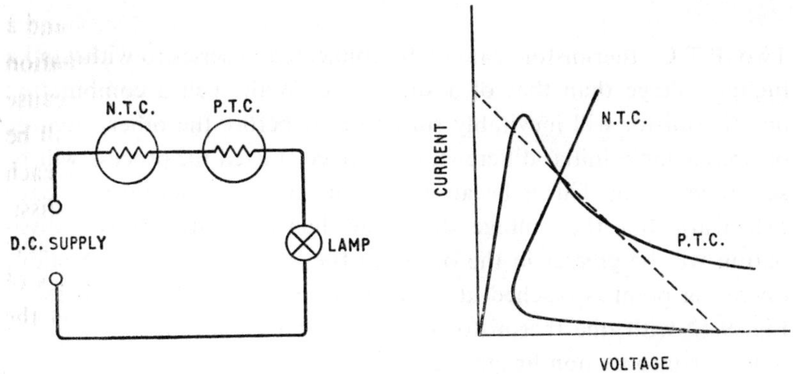

Fig. 9.10. Connection of N.T.C. and P.T.C. thermistors to give relaxation oscillation

frequency-dependent. Characteristics that would not normally lead to oscillation are shown in Fig. 9.11. P_1 is an inherently stable point: oscillation can be induced at P_3, however, by preheating the N.T.C. thermistor or temporarily raising the applied voltage.

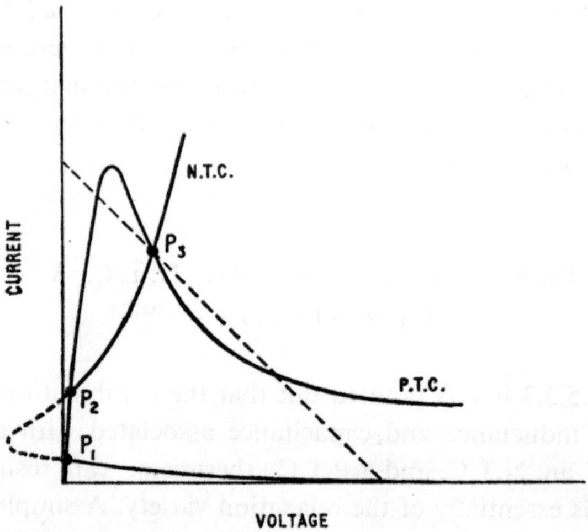

Fig. 9.11. Interconnection of N.T.C. and P.T.C. which is stable

9.7.1. Series and Parallel Operation of Thermistors

Two P.T.C. thermistors cannot be connected in series to withstand a higher voltage than that of a single one. With such a combination one thermistor will inevitably start to heat before the other, because of unavoidable initial differences, and once started the process will be self-augmenting. This is because the same current flows through each thermistor but the voltage drop and hence greater power dissipation will be greater in the one with the higher resistance. A stable operating point is reached at either P_1 or P_2 in the characteristics of Fig. 9.12 with one thermistor operating in its linear region and the other far into its non-linear region.

P.T.C. thermistors can, however, be connected in parallel to give greater current-handling capability, in an inherently stable condition. If one thermistor overheats the colder one then takes a larger fraction of the total current, so the situation is self-regulating.

The opposite is true for N.T.C. thermistors. With a series-connected pair the resistances both tend to a low value, with possible

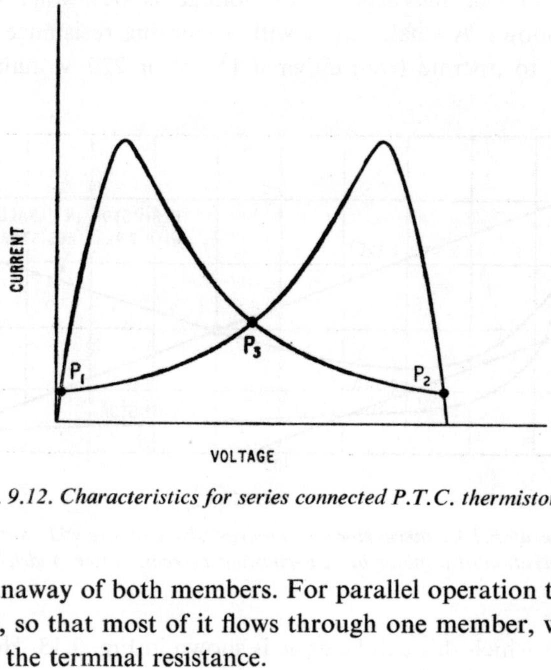

Fig. 9.12. Characteristics for series connected P.T.C. thermistors

thermal runaway of both members. For parallel operation the current will divide, so that most of it flows through one member, which then dominates the terminal resistance.

9.8. SWITCHING APPLICATIONS

Both N.T.C. and P.T.C. thermistors are used in a variety of switching operations in response to current, voltage and thermal overload. Because of their superior temperature coefficient of resistance P.T.C. devices are now most widely used in this area. Andrich[12] has described several applications and these will now be outlined.

9.8.1. Electric Motor Operation at Widely Different Voltages

It is clearly advantageous if a given electric motor can be operated with quite different supply voltages. Since motor current rather than voltage is the quantity which is of real significance, a circuit in which current is almost independent of voltage is desirable. A specific example follows. A small motor with a working resistance of 2·5 kΩ is required to operate from either a 110 V or 220 V mains supply.

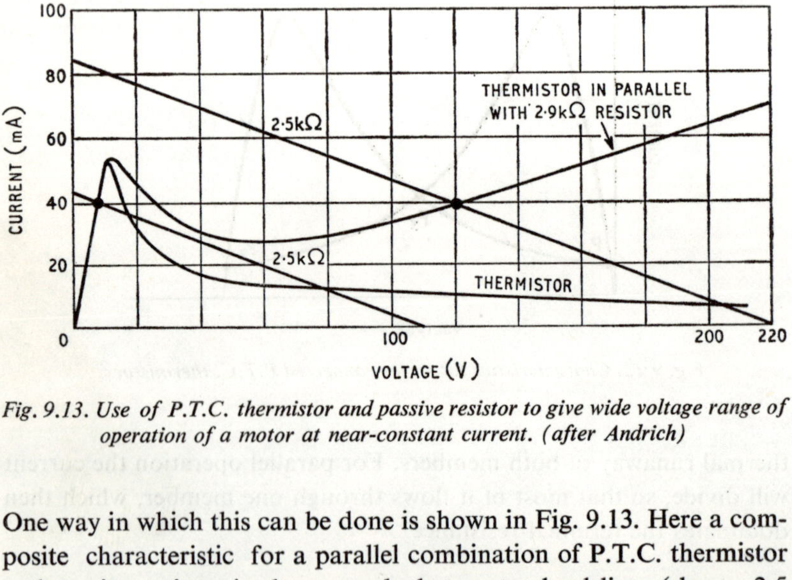

Fig. 9.13. Use of P.T.C. thermistor and passive resistor to give wide voltage range of operation of a motor at near-constant current. (after Andrich)

One way in which this can be done is shown in Fig. 9.13. Here a composite characteristic for a parallel combination of P.T.C. thermistor and passive resistor is chosen such that motor load lines (slope—2·5 kΩ), based on supplies of 110 V and 220 V intersect it once only, at approximately similar values of current. Since there is some delay before the thermistor heats up sufficiently to move the operating point from the linear to the non-linear part of the characteristic, almost the full supply voltage will be applied to the motor for a short time. As a safeguard against this momentary overload and also to compensate for spread and variation of the resistance of the motor it is desirable to connect a V.D.R. in shunt with it.

9.8.2. Motor Protection against Temperature Variation

Small ventilator-cooled electric motors often operate under conditions in which dust accumulates in the ventilator over a long period and often leads to overheating and failure. A safeguard can be provided by connecting a P.T.C. thermistor in series with the motor, mounted in thermal contact with its winding. The load line of the motor normally intersects the thermistor static characteristic in its linear region. If overheating occurs, however, the peak current of the thermistor characteristic is depressed and design is such that this peak falls below the load line so that switching occurs to an intersection at low current and high voltage, before the motor is permanently damaged.

9.8.3. Motor Protection Against Overload

If an electric motor is overloaded or braked while running at its rated voltage then the windings may again be damaged due to Joule heating. Decreasing the speed increases the armature current and this is equivalent to decreasing the effective resistance. This can again be prevented from causing permanent damage by connecting a P.T.C. thermistor in series with the motor, so that the normal operating position is with an intersection of the motor load line with a linear part of the characteristic, and overload, represented by a rotation of the load line clockwise about the applied voltage as a pivot, trips to a high voltage, low current operating point in the non-linear region.

9.8.4. Intermittent Operation of a Power-consuming Device

The design of, for example, an electric motor for intermittent operation can be based on less stringent requirements than for continuous operation, but it is necessary to protect against continuous running for a period of say, more than three minutes. This can be done using an electro-mechanical timer, but use of the switching delay of P.T.C.

thermistor, as described in Section 4.3.3 gives a more simple, compact, and reliable solution. Use of a single thermistor may make it difficult to achieve a breakdown voltage as great as the mains voltage and as long a delay time as three minutes. The delay time can be improved considerably by connecting thermistors in parallel, since this reduces the voltage drop and increases the dissipation constant.

9.8.4. Low-frequency Multivibrator

The instability of series-connected P.T.C. thermistors has been utilised in the circuit shown in Fig. 9.14 to generate relaxation oscillations with a period ranging from 30 to 40 s. The cold resistances of

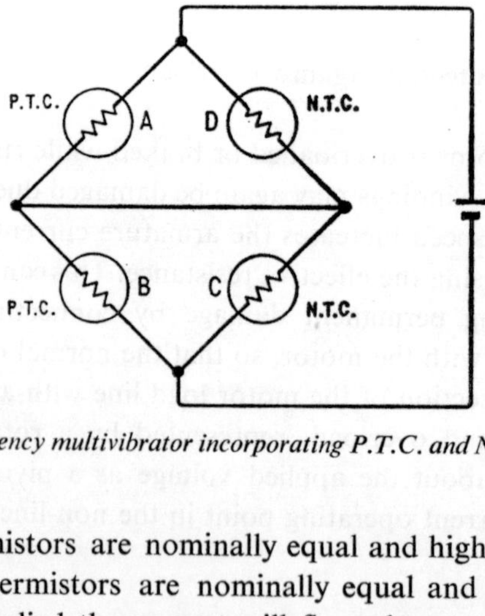

Fig. 9.14. Low-frequency multivibrator incorporating P.T.C. and N.T.C. thermistors

the N.T.C. thermistors are nominally equal and high, whereas those of the P.T.C. thermistors are nominally equal and low. When the voltage is first applied the current will flow almost entirely through arms *A* and *B*. Suppose *A* starts to heat first: as already described this will be regenerative. Eventually the voltage across *A* will rise almost to the applied voltage. This will cause the N.T.C. thermistor, *D*, to be heated and its resistance to fall, again a regenerative process, which progressively diminishes the current in *A* and increases that in P.T.C. thermistor *B*. The resistance of *D* eventually becomes very small and the majority of the applied voltage then appears across *B*. With a proper choice of cold values of resistance *A* will now have returned to its original low resistance value so that as *B*'s resistance approaches its maximum value the current will shift from the path through *B* and *D* to that through *A* and *C* and the cycle then repeats.

REFERENCES

1. MEACHAM, L. A. 'The bridge-stabilised oscillator', *Proc. Inst. Radio. Engrs*, **26**, 1278–94 (1938)
2. SHEPHERD, W. G. and WISE, R. O. 'Variable-frequency bridge-type frequency stabilised oscillators', *Proc. Inst. Radio. Engrs*, **31**, 256–68 (1943)
3. HYDE, F J. 'Highly-selective amplification at low frequencies', *Wireless Engr*, **29**, 85 (1952)
4. FLEMING, L. 'Thermistor-regulated low-frequency oscillator', *Electronics* **19**, 97–9 (1946)
5. WILLIAMSON, R. 'Wide range audio oscillator', *Wireless World*, **62**, 508–9 (1956)
6. MARKUS, J. and ZELUFF, V. 'Thermistor-regulated audio-frequency oscillator' *Handbook of Industrial Electronic Circuits*, **164**, McGraw-Hill (1948)
7. BLEUZE, J. 'Les propriétés et les applications des thermistances', *Onde élect.* **33**, 578–90 (1953)
8. TWEEDDALE, J. E. 'Thermistors', *West. elect. Oscill.*, Dec., 3–5 and 34–7 (1945)
9. JOHNSON, J. C. 'Thermistor technics', *Electron., Inds., N. Y.*, **4**, 74 (1945)
10. BECKER, J. A., GREEN, C. B., and PEARSON, G. L. 'Properties and uses of thermistors-thermally sensitive resistors', *Trans. Am. Inst. elect. Enrgs* **65**, 711 (1946)
11. SABURI, O. and WAKINO, K. 'Processing techniques and applications of positive temperature coefficient thermistors', *I.R.E. Trans. compon. Parts*, **10**, 963 (1963)
12. ANDRICH, E. 'Properties and applications of P.T.C. thermistors', *Electron. Applic.* **26**, 123 (1966)
13. WELLER, J. A. 'Volume-limiter for leased-line service', *Bell Labs Rec.*, **23**, 72 (1945)
14. PATCHETT, G. N. 'Precision a.c. voltage stabilisers', *Electron. Engng.*, **22**, 371, 424, 470, 499 (1950)
15. PATCHETT, G. N. 'The theory of the non-linear bridge circuit as applied to voltage stabilisers', *J. Instn. elect. Engrs.*, **93 Part III**, 16, (1946)
16. EKELÖF, S., BJÖRK, N., and DAVIDSON, R. 'Large-signal behaviour of directly-heated thermistors', *Chalmers tek. Högsk. Handl.*, Nr 185 (1957)
17. GANO, J. J., and SANDY, G. F. 'Thermistors for the gradual application of heater voltages to thermonic tubes', *I.R.E. Trans. electronic Comput.*, **7**, 61 (1958)
18. IMAMURA, E., and TAMADA, M. 'Latest applications of Murata Posistor', *Japan Electron. Engng.*, **12**, 60, (1967)
19. SILLARS, R. W. 'Materials and devices of falling resistance-temperature characteristic, *J. Scient. Instrum.*, **19**, 81 (1942)

20. CAMERON, C. T., and HUMPHRIES, W. A. 'Shared service', *P. O. elect. Engrs'. J.*, **41,** 139 (1948)
21. LINAU, H., and SEIFERT, I. 'Relay time delays with N.T.C. resistors' *Matronics*, **12,** 216 (1957)
22. SCARR, R. W. A., and SETTERINGTON, R. A. 'Thermistors, their theory, manufacture and application', *Proc. Inst. elect. Engrs.*, **107B,** 395, 408 (1960)
23. HYDE, F. J. 'Impedance of a thermistor at low frequencies', *J. Electron.*, **3,** 303 (1955)
24. BROWNLIE, J. D. 'Thermal electric oscillators', *J. Instn. elect. Engrs.*, **5,** 532 (1959)

CHAPTER 10

Indirectly-heated Thermistors

10.1. INTRODUCTION

In previous chapters particular attention has been paid to two-terminal, directly-heated thermistors, whose temperatures have been determined by the combined effects of the ambient temperature and of the electrical power dissipated in the thermistor bead. In the indirectly-heated thermistor[1] (see Section 1.3.1) the thermistor bead is not generally connected directly into a primary control circuit. Its temperature is governed by the ambient temperature, the power dissipated in the heater winding, to which it is thermally, but not electrically connected, and the power dissipation in the bead itself. A convenient symbol is shown in Fig. 10.1, in which R is the semiconductor resistance and R_h the heater resistance.

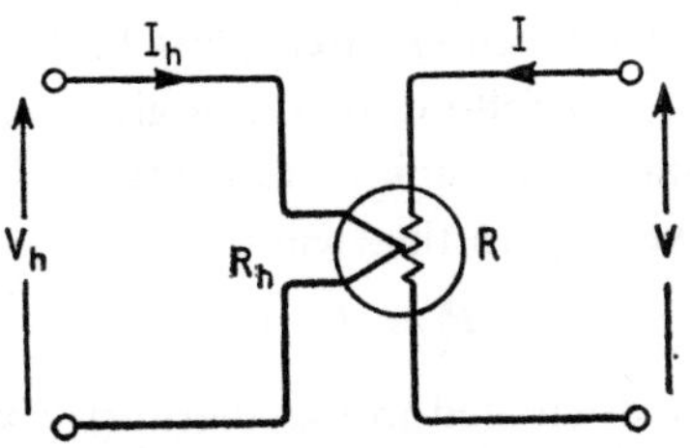

Fig. 10.1. Symbol for indirectly-heated thermistor

Typical current-voltage characteristics for an N.T.C. indirectly-heated thermistor are shown in Fig. 1.5; here heater current is the parameter and the bead current is constant. Corresponding curves relating bead resistance to bead power with heater power as parameter are also used. As the heater resistance increases with temperature, similar curves to those referred to above, but with heater power as parameter, will differ somewhat. Generally, however, the variation in R_h is so small that it can be neglected in practice. The difference between characteristics based on a constant or temperature-dependent R_h differ by less than $\pm 1\%$.

The electrical properties of the bead itself are similar to those of directly-heated thermistors. In transient and a.c. applications, account has to be taken of the additional delay introduced by the thermal inertia of the heater, which is usually a winding having a resistance of about 100 Ω. This is generally incorporated in a single 'thermal time constant'. Most commercially-available types are limited to operation in the audio frequency range or below, by the relatively large inductance of the heater winding. Special r.f. types have been made[2], which have a lead inductance of less than 0·5 μH.

Operation at the lower end of the frequency spectrum is determined by the thermal time-constant of the thermistor. For type-B thermistors this is about 18 s, but for miniature r.f. types it is of the order of 1 s.

10.2. STATIC CONDITIONS

From the theoretical standpoint it is useful to assume isothermal conditions at temperature T, throughout the bead, insulating-cover and that part of the heater which is in intimate contact with it, and that the remaining parts (connecting wires, glass bulbs etc.) are at the ambient temperature, T_a. It follows that at a given ambient temperature the thermistor resistance is a unique function of T, and in turn of the total dissipated power, P_t. In the steady state

$$P_t = P + \eta P_h \tag{10.1}$$

Here P is the power dissipated in the thermistor bead and P_h is that dissipated in the heater. η is the heater-efficiency. This accounts for the fact that some of the total heater power is lost in parts of the heater not in contact with the thermistor bead. η can be computed from pairs of values of P_1, P_{h1} and P_2, P_{h2} corresponding to the same value of P_t and hence of R, as follows:

$$\eta = -\frac{(P_2 - P_1)}{(P_{h2} - P_{h1})} \tag{10.2}$$

In the limit

$$\eta = -\left(\frac{\mathrm{d}P}{\mathrm{d}P_h}\right)_{R=\mathrm{constant}} \tag{10.3}$$

Since it has been found[3] that η is practically constant it is not necessary to determine the complete V–I or R–P characteristics, through time-consuming direct measurements. It is sufficient to determine η and the R–P characteristic for $P_h = 0$. For precise work, account must be taken of the increase of heater resistance with temperature: as this is approximately linear there is no serious problem.

10.2.1. The d.c. Resistance Equation

Since R is a unique function of T this may be written, following Equation 2.7, as

$$R = R_c \exp\left\{\frac{B}{(P+\eta P_h)/K+T_a}\right\} \tag{10.4}$$

i.e. the heater power P_h has the same influence on the static characteristic as an increase in ambient temperature, by an amount $\eta P_h/K$. The constants B, R_c and K can be determined in the way described in Section 3.2.7.

10.3. DYNAMIC CONDITIONS

If Newtonian cooling is assumed, the powerbalance equation for the thermistor body is

$$P_r = P + \eta P_h = K\theta + C\frac{\mathrm{d}\theta}{\mathrm{d}t} \tag{10.5}$$

K, C and θ have the usual meaning and $C/K=\tau$ is the thermal time-constant. For analytical purposes it is convenient to disregard the slight decrease of τ with $K\theta$.

Dynamic problems can be solved using Equation 10·5. If $P \approx 0$ then the simpler form results:

$$\eta P_h = K\theta + \tau \frac{d(K\theta)}{dt} \tag{10.6}$$

With $P_h(t)$ known, $K\theta$ can be obtained by direct integration and $R(t)$ can then be found from the R–P characteristic. For example, if P_h is suddenly changed from P_{h1} to P_{h2}, then

$$K\theta(t) = \eta P_{h1} + \eta(P_{h2} - P_{h1})\{1 - \exp(-t/\tau)\} \tag{10.7}$$

10.3.1. Small-signal Equivalent Circuit

The derivation of a small-signal equivalent circuit for both heater and bead circuits was carried out by Björk[3], and his approach will be followed here. It is assumed that the quantities involved are expressed as average values, plus instantaneous deviations from these, as follows:

$$\begin{aligned} V &= V_0 + \Delta V \qquad & P &= P_0 + \Delta P \\ I &= I_0 + \Delta I \qquad & P_h &= P_{h0} + \Delta P_h \\ R &= \bar{R} + \Delta R \qquad & P_t &= P_{t0} + \Delta P_t \end{aligned}$$

with $\bar{R}^* = V_0/I_0$ and $P_0 = V_0 I_0$. With appropriate substitutions in Equation 10.5 we obtain

$$\Delta P + \eta\,\Delta P_h = \Delta(K\theta) + \tau \frac{d}{dt}(\Delta(K\theta)) \tag{10.9}$$

Assuming $\Delta R \ll \bar{R}$ and using $\Delta P_t = (1/m)\Delta R/\bar{R}$, where m is the power coefficient of bead resistance, this can be written as

$$\Delta P + \eta\,\Delta P_h = \frac{1}{m\bar{R}}\left\{\Delta R + \tau \frac{d}{dt}(\Delta R)\right\} \tag{10.10}$$

* $\bar{R}$ is used here since R_0 is used elsewhere to represent the zero frequency slope resistance. In fact $\bar{R} = R_\infty$, the infinite frequency slope resistance.

It is now assumed that variations in current and voltage on the thermistor-bead side are small, so that

$$\left.\begin{aligned}\Delta P &= P_0\left(\frac{\Delta V}{V_0}+\frac{\Delta I}{I_0}\right)\\ \Delta R &= \bar{R}\left(\frac{\Delta V}{V_0}-\frac{\Delta I}{I_0}\right)\end{aligned}\right\} \qquad (10.11)$$

Introduction of Equations 10.11 into 10.10 and replacement of $\mathrm{d}/\mathrm{d}t$ by $\mathrm{j}\omega$, ΔV by v and ΔI by i gives

$$P_0\left(\frac{v}{V_0}+\frac{i}{I_0}\right)+\eta\,\Delta P_h = \frac{i}{m}\left(\frac{v}{V_0}-\frac{i}{I_0}\right)(1+\mathrm{j}\omega\tau) \qquad (10.12)$$

Defining $F = -mP_0$ and rearranging, leads to

$$v - i\bar{R}\frac{(1-F+\mathrm{j}\omega\tau)}{(1+F+\mathrm{j}\omega\tau)}+\frac{F}{I_0}\,\frac{\eta\,\Delta P_h}{(1+F+\mathrm{j}\omega\tau)} = 0 \qquad (10.13)$$

This is satisfied by the equivalent circuit of Fig. 10.2a if

$$Z(\omega) = \bar{R}\left(\frac{1-F+\mathrm{j}\omega\tau}{1+F+\mathrm{j}\omega\tau}\right) \qquad (10.14)$$

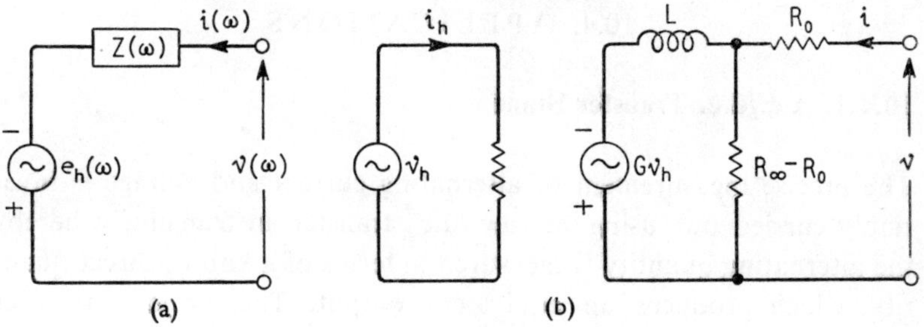

Fig. 10.2. Equivalent circuits of indirectly-heated N.T.C. thermistor; (a) Thévenin circuit; (b) circuit containing individual lumped circuit elements

and

$$\mathrm{e}_h(\omega) = \frac{F}{I_0}\,\frac{\eta\,\Delta P_h(\omega)}{(1+F+\mathrm{j}\omega\tau)} \qquad (10.15)$$

An equivalent form incorporating lumped components and a generator $Gv_h(t)$ to represent heater voltage variations as shown in Fig.

10.2b. Here

$$R_0 = R_\infty \frac{1-F}{1+F}$$

$$L = R_\infty \tau \frac{2F}{(1+F)^2}$$

$$G = \frac{\eta I_{h0}}{I_0} \cdot \frac{2F}{1+F}$$

R_∞, ($\bar{R}$), is the slope of the straight line from the origin intersecting the operating point, defined by (I_0, V_0), and R_0 is the slope resistance of the static characteristic at this point, as defined earlier for two-terminal thermistors. If the heater current is constant $v_h(t) = 0$.

Björk made measurements of the impedance of a Stantel thermistor, type B/2321/60, with different values of P_h and P and obtained curves approximating the semicircles defined by the equivalent circuit of Fig. 10.2b. The agreement was not as good as has been obtained for two-terminal thermistors.[4] This is almost certainly due to the oversimplification made concerning temperature distribution in the present case.

10.4. APPLICATIONS

10.4.1. A.c./d.c. Transfer Standard

The precise measurement of alternating current and voltage is invariably carried out using an a.c./d.c. transfer instrument, whereby the alternating quantity is measured in terms of a known direct quantity, which produces an equivalent output. The a.c./d.c. transfer devices in general use are (a) the electrostatic voltmeter which can be used up to about 20 kHz; (b) the reflecting electro-dynamic instrument, which requires a relatively high power consumption and is restricted to low audio frequencies; (c) the thermal convertor, using a heated thermocouple in vacuum—this has been investigated in great detail and the errors have been carefully formulated—the sensitivity is relatively low, the the output being about 6 mV across a few ohms at full rated heater current and for a definition of 1 part in 10^4 the galvanometer used must be capable of detecting 0·5 μV; special thermocouples have been used at v.h.f. Widdis[5] has considered the use of the indirectly-heated thermistor as an a.c./d.c. transfer standard, with the assumption that self-heating in the thermistor bead is negligible. Complementary work has been done by Schrader[6]. Widdis used thermistors type B5412/60 in the bridge-circuit shown in Fig. 10.3. Here R_1 is a fixed resistor and R_2 can be varied over a limited range for initial balancing purposes, with the same a.c. and d.c. fed to both heaters, depending on use as a standardising device or for the actual measurement of a given alternating current. The bridge supply is 9 V and the detector is a portable reflecting galvanometer. The operating current range for the heaters is between 2·5 and 15 mA.

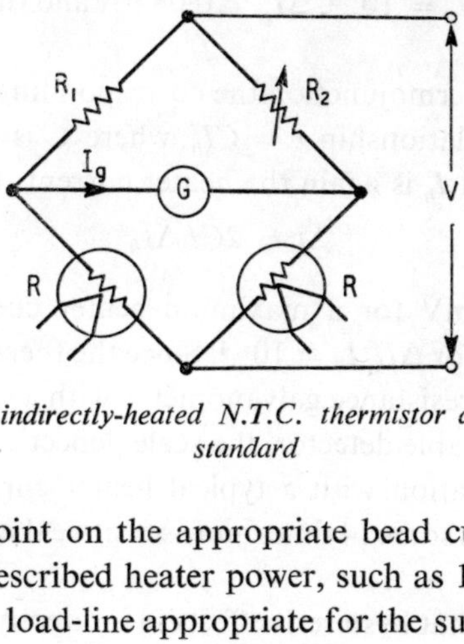

Fig. 10.3. Use of an indirectly-heated N.T.C. thermistor as an a.c./d.c. transfer standard

The operating point on the appropriate bead current-voltage characteristic, for a prescribed heater power, such as 10 mW, can readily be found using the load-line appropriate for the supply voltage to the bridge and the passive series resistance arm.

An expression for bridge sensitivity for an equal-ratio-arm bridge is obtained by determining the change in detector current I_g associated with a change ΔR in the resistance of one of the thermistor bead resistances, produced by a change ΔI_h in heater current. From Fig. 10.3 it follows that

$$\Delta I_g = \frac{VR_1\,\Delta R}{R_g(R_1+R)^2+2RR_1(R_1+R)+\Delta R\{R_1^2+2RR_1+R_g(R_1+R)\}} \tag{10.17}$$

By definition, $\alpha = (1/R)(dR/d\theta)$, so that $\Delta R = \alpha R \, \Delta T$, where $\Delta T = \Delta P/K$, ΔP being given by $2IR_h \, \Delta I_h$, if the variation of R_h with I_h is negligible. Hence $\Delta R = 2\alpha I_h R R_h \, \Delta I_h/K$, which, when substituted into Equation 10.17, with neglect of terms in ΔR in the denominator, gives

$$\Delta I_g = \frac{VR_1R}{R_g(R_1+R)^2+2RR_1(R_1+R)} \cdot \frac{2I_h\alpha R_h \, \Delta I_h}{K} \tag{10.18}$$

α varies with the operating current, but for an approximate calculation of sensitivity can be taken as 0·3.

Taking the sensitivity of the galvanometer as 250 mm/μA, its internal resistance R_g as 450 Ω and $R_1 = 10$ kΩ, $R = 6{\cdot}0$ kΩ, $R_h = 100$ Ω, $K = 0{\cdot}2 \times 10^{-3}$ mW/°C, with $I_h = 10$ mA, it follows from Equation 10.18 that for $\Delta I_h/I_h = 10^{-4}$, $\Delta I_g = 0{\cdot}08$ μA and the scale deflection is 20 mm.

For a vacuum thermojunction the corresponding sensitivity is determined from the relationship $v = CI_h^2$, where C is a constant, v is the output voltage and I_h is again the heater current. It follows that

$$\Delta v = 2CI \, \Delta I_h$$

Typically $v = 6{\cdot}0$ mV for a maximum heater current of 10 mA, so that $\Delta v = 1{\cdot}2$ μV for $\Delta I_h/I_h = 10^{-4}$. Since the thermocouple resistance is 9 Ω and a 10 Ω resistance galvanometer with a sensitivity of 23 mm/μA would be a suitable detector, the scale deflection would be 1·4 mm.

Hence, for operation with a typical heater current of 10 mA, the thermistor bridge is some 14 times more sensitive than the vacuo-thermojunction.

Widdis showed that hysteresis effects in indirectly-heated thermistors of the type used by him were negligible for heater power dissipations less than 10 mW. Because of their relatively long thermal time constants they could be used to frequencies as low as 0·2 Hz which is well below the working range of most other transfer instruments, and up to a frequency of 500 kHz with a departure from the middle frequency response of 1 part in 10^4. The response time is rather slow and starting with the thermistors in a cold condition the time required to achieve balance to an equivalent heater-current difference of 1 part in 10^4 is theoretically about 80 s but in practice considerably more.

10.4.2. Radio Frequency Milliammeter

Wood[2] has shown that specially-made indirectly-heated thermistors can be used in a similar application for the measurement of radio frequency currents with an accuracy of ±2% at 7 MHz, for a full-scale deflection of 10 mA r.m.s. A bridge circuit was again used. Radio frequency current applied to one N.T.C. thermistor reduces the resistance of its bead. This causes a bridge unbalance which is amplified and fed to the heater of the second thermistor in the adjacent arm, whose resistance is thereby reduced to maintain balance. If the gain of the d.c. amplifier is very large then approximately equal r.m.s. currents flow in both heaters and the alternating current is measured in terms of the direct current supplied to the heater of the bridge-fed thermistor.

10.4.3. Low-frequency oscillation

The use of directly-heated thermistors in sub-audio frequency oscillator circuits has been described in Section 5.3.3. Indirectly-heated thermistors can also be used for this purpose[3,7] A basic circuit incorporating an N.T.C. thermistor is shown in Fig. 10.4. In this

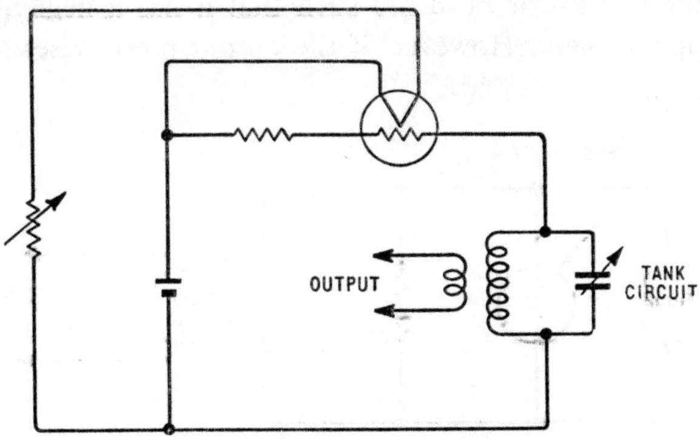

Fig. 10.4. Indirectly-heated N.T.C. thermistor oscillator circuit

particular case the negative a.c. resistance of the thermistor bead, which is controlled by varying the heater current, is effectively applied in parallel with an *L–C* tank circuit. It has been claimed that oscillation up to 5 KHz has been achieved in 'very small' thermistors having correspondingly small thermal time constants.

10.4.4. Gain control

The application of directly-heated thermistors to amplifier gain control has been considered in Section 9.3. Indirectly-heated N.T.C. thermistors can also be used for this purpose. A basic circuit used in the Bell type K2 carrier system[8] is shown in Fig. 10.5. This incorporates a compensating circuit in which an N.T.C. disc thermistor is used to offset changes in ambient temperature by controlling the current in the heater feed-circuit. Hence the operating resistance of the thermistor bead is principally determined by the circuit in which it is connected (see Section 9.3).

Another type of thermistor gain control system[10] incorporates an indirectly heated thermistor in which the heater is driven by the output of the amplifier and with the thermistor bead connected in shunt with the input circuit, as shown in Fig. 10.6. Here the feedback is of a thermal rather than electrical nature. The same circuit gives protection against gain overloads. For normal signals the sensitivity and resistance of the thermistor bead are such that it has a negligible effect on the amplifier gain. However, if the output power rises to an ab-

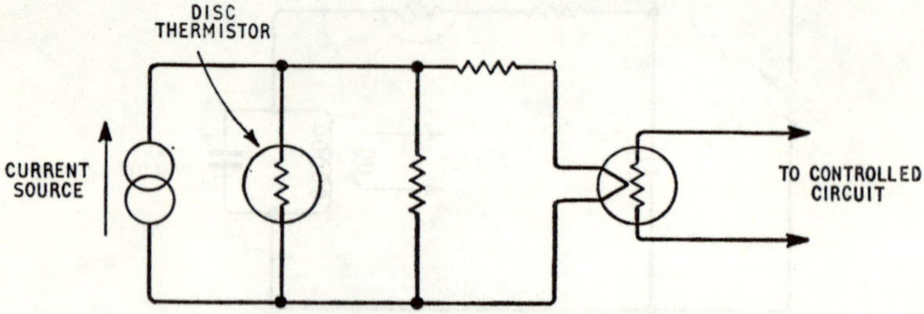

Fig. 10.5. Basic control circuit of Bell Laboratories K2 system

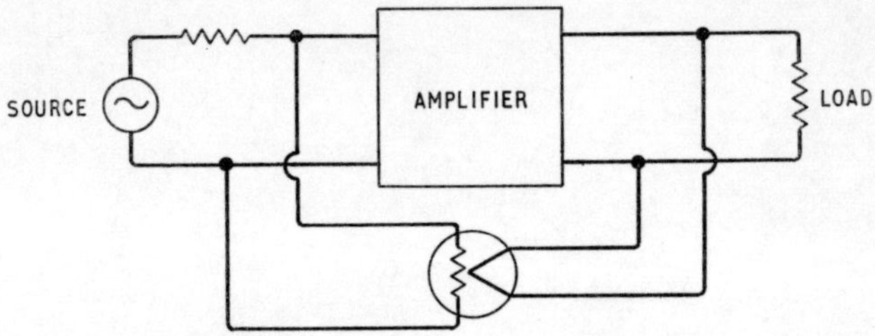

Fig. 10.6. Gain control with an indirectly-heated thermistor in the feedback loop

normal level, the thermistor becomes heated and the bead resistance falls so that it shunts the amplifier input, thereby limiting the input to it and offsetting the abnormal power output.

REFERENCES

1. BECKER, J. A., GREEN, C. B., and PEARSON, G. L. 'Properties and uses of thermistors–thermally sensitive resistors', *Trans. Am. Inst. elect. Engrs.*, **65**, 711 (1946)
2. WOOD, H. B. 'An r.m.s. milliammeter of novel design for the measurement of current from zero to video frequencies', *J. Scient. Instrum.* **31**, 124 (1954)
3. BJÖRK, N. 'Theory of the indirectly-heated thermistor', *Chalmers tek. Högsk. Handl.*, No. 211 (1959)
4. HYDE, F. J. 'Impedance of a thermistor at low frequencies', *J. Electronic.* **1**, 303 (1965)
5. WIDDIS, F. C. 'The indirectly-heated thermistor as a precise a.c./d.c. transfer device', *Proc. Instn. elect. Engrs.*, **103B**, 693 (1956)
6. SCHRADER, H. J. 'A.C. compensation measurement with indirectly-heated thermistors', *Elektrotech Z.*, **73A**, 547 (1952)
7. ROSENBERG, W. 'Thermistors', *Electronic Engng.*, **19**, 185 (1947)
8. BROOKS, F. A. 'The K2 carrier system', *Bell Labs Rec.*, **27**, 141 (1949)
9. BELLMAN, J. H. 'A pilot-channel regulator for the K1 carrier system', *Bell Labs Rec.* **20**, 258 (1942)
10. JOHNSON, J. C. 'Thermistor Technics', *Electronic Inds.* **4**, 74 (1945)

Index